WATER SUPPLY BYELAWS GUIDE
Second Edition

Editors:
S. F. WHITE
G. D. MAYS
WRc

Routledge
Taylor & Francis Group

LONDON AND NEW YORK

First published in 1986 by WRc Evaluation and Testing Centre
This Second Edition first published in 1989

Published 2014 by Routledge
2 Park Square, Milton Park, Abingdon, Oxon OX14 4RN
711 Third Avenue, New York, NY 10017, USA

Routledge is an imprint of the Taylor & Francis Group, an informa business

ISBN13: 978-0-13-950395-5 (pbk)

British Library Cataloguing in Publication Data
Water supply byelaws guide. 2nd edition.
1. Water — Law and legislation — England
I. White, S. F.   II. Mays, G. D.   III. WRc
344.203'92'02636   KD2555

GUIDE TO THE APPLICATION AND INTERPRETATION OF WATER BYELAWS

CONTENTS

This Guide is produced under the auspices of Water Research and is based on an original draft by Water Training.

FOREWORD

1. All the organisations responsible for water supply - the 10 water authorities and the 28 water companies in England and Wales, the 12 Regional and Islands Councils in Scotland and the Department of the Environment in Northern Ireland - have the statutory power to make, and the duty to enforce, byelaws (regulations in Northern Ireland) for the prevention of waste, undue consumption, misuse or contamination of water supplied by them. The Model Water Byelaws 1986 Edition formed the basis for such byelaws in Great Britain and for regulations in Northern Ireland. These have now been made.

2. In February 1985 a Byelaws Guidance Panel(1) was formed by the United Kingdom Water Fittings Byelaws Scheme(2) to produce a Guide to Byelaws the aim being that it should:

   (a) assist water inspectors, manufacturers, architects, consultants, merchants, installers and consumers to understand byelaws based on the Model;

   (b) bring about uniformity of approach to byelaw enforcement throughout the UK;

   (c) be suitable for use as a training document;

   (d) be subject to continual review by the Panel to ensure its compliance with the latest requirements;

   (e) comply with byelaw guidance issued by the Department of the Environment; and

   (f) be based wherever possible on British Standards.

3. This Guide is intended to be applicable throughout the United Kingdom and includes a copy of each byelaw. Following each byelaw, (or section of byelaw), there is guidance and illustrations where it is considered that they may be helpful. While it should be a reliable source of guidance in most cases, it must be emphasised that the Guide has no legal force or standing. It is always a matter for the enforcing undertakers to be satisfied that any interpretation, whether based on the Guide or otherwise, meets the requirements of their byelaws.

4. Compliance with byelaws implies no particular degree of fitness for purpose of components used in water services. Fitness for purpose

can only be achieved by the use of products made, quality assured and installed to recognised specifications such as British Standards. So wherever possible, appropriate British Standards are cited in the Guide. The inclusion of such references implies that products which comply with such standards will meet the relevant requirements of byelaws. The suitability of British Standards is kept constantly under review and they are amended or reissued as circumstances require. A list of such standards is included in Appendix 3.

5. Amendments to the Guide will be notified to the technical press and elsewhere.

    All enquiries concerning the content of the Guide should be addressed to Water Byelaws Advisory Service, 660 Ajax Avenue, Slough, Berkshire SL1 4BG (Tel: (0753) 37277).

6. All enquiries concerning training for those whose work requires an understanding of the byelaws, should be addressed to Water Training(3), Development Section, Tadley Court, Tadley Common Road, Tadley, Basingstoke, Hampshire RG26 6TB (Tel: (07356) 3011).

7. **Disclaimer**

    Neither the UK Water Industry nor the producers of this guide accept any responsibility for any acts or omissions on the part of any person or body designing, installing or inspecting water supply systems which may arise from the use of this Guide.

(1) For constitution of Panel producing Guide, see Appendix A.

(2) The UK Water Fittings Byelaws Scheme makes provision throughout the UK for the assessment and testing of water fittings to ascertain whether they comply with byelaws. The Scheme is serviced by the Water Byelaws Advisory Service (WBAS) of the WRc situated at Slough.

(3) Water Training was set up by the United Kingdom Water Industry to provide it with direct training services. Courses designed for non-Water Industry organisations are also available both in residential training centres and on-site.

NOTES ON THE USE OF THE GUIDE

1. Using the guide

Each byelaw is indexed by number and is distinguished from the following guidance by a tone background. This guidance contains information and/or diagrams relevant to that particular byelaw where it is considered that it would be helpful.

2. Guide references

These references, at the end of the book, list, in alphabetical order, subject headings which identify major components, locations, or physical condition of a water installation in relation to byelaw numbers.

3. Dimensions

Generally, metric dimensions are quoted in byelaws. Conversion tables are included in Appendix B.

4. British Standards

Reference to British Standards in the Guide exclude the year of publication and are intended to be the current versions including amendments thereto. For brevity the words "Specification for" have been omitted from the titles in this book.

5. Codes of Practice

Particular attention is drawn to British Standard Specification 6700: Design, installation, testing and maintenance of services supplying water for domestic use within buildings and their curtilages. This is a code of practice replacing CP 310 and CP 99 and throughout this Guide is referred to as BS 6700.

6. Accepted fittings

Classified lists of accepted fittings and materials have not been included in this guide as they are available from the Water Byelaws Advisory Service, 660 Ajax Avenue, Slough, Berkshire SL1 4BG (Tel: (0753) 37277).

The published list, updated every six months, is entitled "Water Fittings and Materials Directory", hereafter referred to as "the Directory". Installation requirements for certain types of fittings are indicated in the Directory and form part of their acceptance.

7. **Diagrams**

The diagrams show only the detail necessary to illustrate points in the accompanying text; they are not intended to be installation diagrams nor do they necessarily include any other byelaw requirement.

8. **Registered Plumbers**

Those entered on the Register of Plumbers maintained and supervised by the Institute of Plumbing are identifiable by the title Registered Plumber and the designatory letters "RP" after their name.

9. **Key to symbols**

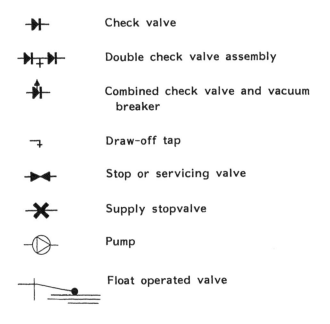

Check valve

Double check valve assembly

Combined check valve and vacuum breaker

Draw-off tap

Stop or servicing valve

Supply stopvalve

Pump

Float operated valve

Diagram 1

INDEX TO BYELAWS

x

xii

PART XIV - REVOCATION

Byelaw

Part I

GENERAL PROVISIONS

CONTENTS

These byelaws are made by *
(hereinafter called "the undertakers") under **
for preventing waste, undue consumption, misuse or
contamination of water supplied by them.

* The name and address of the Water Authority, Company, or Regional and Island Council would be inserted here.

** Here insert

"Section 17 of the Water Act 1945", for undertakers operating in England or Wales; or

"Section 70 of the Water (Scotland) Act 1980", for undertakers in Scotland.

Regulations are made in Northern Ireland for the same purposes as byelaws in the rest of the UK and generally conform with them.

PART I - GENERAL PROVISIONS
BYELAW 1 - INTERPRETATIONS

The General Provisions of the byelaws contain a number of interpretations of terms used. These are not exhaustive and generally interpretations of other terms set out in BS 4118: Glossary of sanitation terms, are acceptable. Where there is any divergence between the byelaw interpretation and that in BS 4118 the former will prevail.

The following are the byelaw interpretations with some interspersed notes and diagrams.

In these byelaws, unless the context otherwise requires:

BACKFLOW means flow in a direction contrary to the intended normal direction of flow;

BACKSIPHONAGE means backflow caused by the siphonage of liquid from a cistern or appliance into the pipe feeding it;

BOILER means an enclosed vessel in which water is heated by the direct application of heat;

CISTERN means a fixed container for holding water at atmospheric pressure;

CLOSED CIRCUIT means any system of pipes and other water fittings through which water circulates but from which no water is drawn for use, and includes any vent pipe fitted thereto but not the feed cistern or the cold feed pipe;

Water evaporated in a cooling system is regarded as drawn-off for use and such a system is not a closed circuit.

COMBINED FEED AND EXPANSION CISTERN means a cistern for supplying cold water to a hot water system without a separate expansion cistern;

COMMUNICATION PIPE means that part of a service pipe which is vested in the undertakers;

CYLINDER means a cylindrical closed vessel capable of containing water under pressure greater than atmospheric pressure;

DISTRIBUTING PIPE means any pipe (other than an overflow pipe or a flush pipe) conveying water from a storage cistern, or from a hot water apparatus supplied from a feed cistern, and under pressure from that cistern;

The word "distributing" is used because the pipe is distributing water from a storage cistern or cylinder to one or more fittings/appliances. A distributing pipe may carry either hot or cold water (see diagram 6 of water service under definition of water fittings).

DOUBLE FEED INDIRECT CYLINDER means an indirect cylinder which has separate cold feed pipe connections for both the primary circuit and the secondary circuit;

EXPANSION CISTERN means a cistern connected to a water heating system which accommodates the increase in volume of water in that system when it is heated from cold;

FEED CISTERN means any storage cistern used for supplying cold water to a hot water apparatus, cylinder or tank;

Diagram 2 gives examples of feed cisterns.

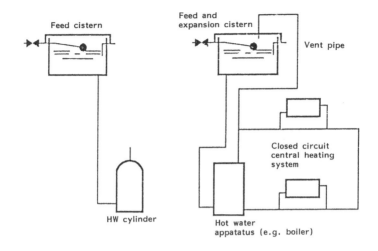

Diagram 2 - Examples of feed cisterns

FLOAT-OPERATED VALVE means a valve, for controlling the flow of water into a cistern, the valve being operated by the vertical movement of a float riding on the surface of the water;

FLUSHING CISTERN means a cistern provided with a device for discharging the stored water rapidly into a watercloset pan or urinal;

FLUSH PIPE means a pipe for conveying water from a flushing cistern to a watercloset pan or urinal;

FLUSHING TROUGH means a flushing apparatus which combines several discharging units in one long cistern body to allow more frequent flushing of two or more watercloset pans;

INDIRECT CYLINDER means a hot water cylinder in which the stored water is heated by a primary heater through which hot water is circulated from a boiler or gas circulator without mixing of the primary and secondary water taking place;

INSTANTANEOUS WATER HEATER means an appliance in which water is immediately heated as it passes through the appliance;

PRIMARY CIRCUIT means an assembly of pipes and fittings in which water circulates between a boiler or other water heater and the primary heater inside a hot water storage vessel;

5

PRIMARY HEATER means a heater mounted inside a hot water storage vessel for the transfer of heat to the stored water from circulating hot water;

SECONDARY CIRCUIT means an assembly of pipes and fittings in which water circulates in distributing pipes to and from a water storage vessel;

SECONDARY SYSTEM means that part of a hot water system comprising the cold feed pipe, any storage cistern, water heater and flow and return pipework from which hot water for use is conveyed to all points of draw-off;

SERVICE PIPE means so much of any pipe for supplying water from a main to any premises as is subject to water pressure from that main, or would be so subject but for the closing of some valve;

Diagram 3 illustrates the terms "service pipe", "supply pipe" and "communication pipe".

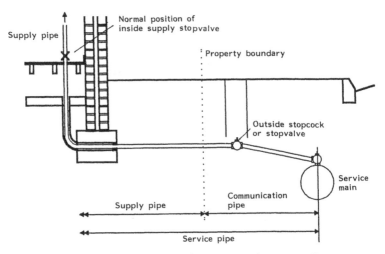

Diagram 3 - Typical domestic water service connection

6

SERVICING VALVE means a valve for shutting off the flow of water in a pipe connected to a water fitting to facilitate the maintenance or servicing of that fitting;

SINGLE FEED INDIRECT CYLINDER means an indirect cylinder which has only one cold feed pipe connection to supply both the primary and secondary waters, so designed that the formation of an air seal during filling prevents mixing and accommodates expansion of the primary water;

SPILL-OVER LEVEL means the level at which the water in a cistern or vessel will first spill over if the inflow exceeds the outflow through any outlet and any overflow pipe;

STOPVALVE means a valve, other than a servicing valve, fitted in a pipeline for controlling or stopping at will, the flow of water;

Certain stopvalves are required to be fitted under byelaws 62 and 63. These are sometimes referred to as "statutory", "inside" or "supply" stopvalves although the terms have no legal significance.

STORAGE CISTERN means any cistern storing water for subsequent use, other than a flushing cistern;

A storage cistern may supply cold water to either cold water fittings/apparatus, or to cold and hot water fittings/apparatus, but if the cistern supplies only hot water apparatus (e.g. a cylinder from which water is drawn for domestic purposes), although technically it is still a storage cistern, a more accurate description is "feed cistern" (see earlier definition). Note also that the definition includes any storage cistern, e.g. in an industrial establishment, where water is stored.

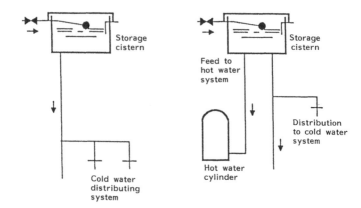

Diagram 4 - Examples of storage cisterns

Examples of vent pipes are shown on diagram 6 under the definition of water fittings. Vent pipes are not only used in connection with escape of air but they also limit pressure rises in installations.

An example of a warning pipe discharging in a conspicuous position is illustrated in diagram 5.

8

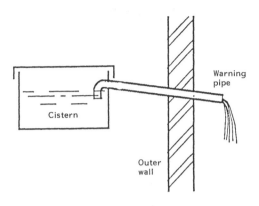

Warning
pipe

Cistern

Outer
wall

Diagram 5 - Discharge beyond an outer wall

WASHING TROUGH means a wash basin, wash trough or sink measuring internally more than 1.2m over its longest or widest part, at which two or more persons can wash at the same time;

WATER SUPPLIED FOR DOMESTIC PURPOSES means water supplied by the undertakers for drinking, washing, cooking and sanitary purposes and includes, where water is drawn from a tap inside a dwelling and no hosepipe or similar apparatus is used, watering a garden and washing vehicles kept for private use;

Domestic purposes includes water taken for the purposes of drinking, washing, cooking and sanitary purposes within an industrial or commercial premises. The definition precludes the supply of water for watering a garden or washing a vehicle by means of a hosepipe as being water for domestic purposes. Any consumer who wishes to use a hosepipe should first seek the advice of the undertakers.

WATER SUPPLIED FOR NON-DOMESTIC PURPOSES means water supplied by the undertakers for domestic purposes which has been drawn off for use, water which is unfit for human consumption and water from any source other than the undertakers' mains;

"Mains" in this definition means pipes laid by the undertakers for the purpose of giving a general supply of water as distinct from a supply to individual consumers.

Water taken for cooling purposes is included in this definition as water that has been drawn off and used whether for evaporative cooling or not.

WATER FITTINGS includes pipes (other than mains), taps, cocks, valves, ferrules, meters, cisterns, baths, waterclosets, soil pans and other similar apparatus used in connection with the supply and use of water;

The general term "water fittings" is used to avoid repeating a list of separate fittings each time and will include any new fitting which may be introduced. The definition makes reference not only to apparatus used in conjunction with the conveyance of water but also connected appliances, e.g. baths, etc.

Most of the fittings used are shown in diagram 6 but in addition they include garden hosepipes, backflow prevention devices, baths, bidets, dishwashers, etc.

In many parts of the United Kingdom all supplies to cold water taps and WC flushing cisterns, etc. are connected directly to the supply pipe. Such connections, under pressure from the mains, are also part of the supply pipe.

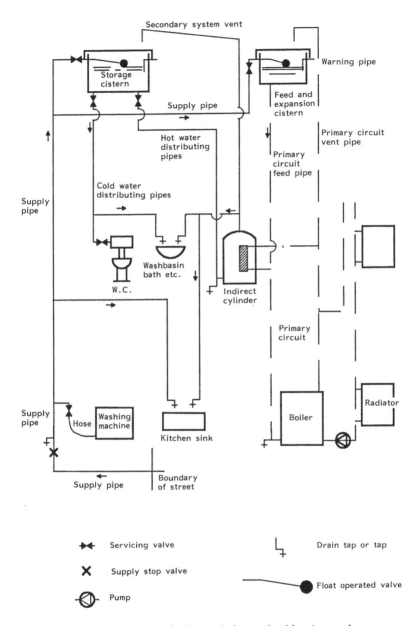

Diagram 6 - Example of a domestic hot and cold water service

11

BYELAWS 2 AND 3

GENERAL PROHIBITIONS

BYELAW 2. No person shall -

    (a) install a water fitting to convey or receive water supplied by the undertakers, or alter, disconnect, or use such a water fitting; or

    (b) cause, or permit such a water fitting to be installed, altered, disconnected or used;

    in contravention of any of these byelaws.

BYELAW 3. No person shall -

    (a) install or cause or permit to be connected or arranged a water fitting in a manner which; or

    (b) use, or cause or permit to be connected or used, a water fitting which is damaged, worn or otherwise faulty so that it;

    causes, or is likely to cause, waste, undue consumption, misuse or contamination of water supplied by the undertakers.

GUIDANCE BYELAW 2

Water fittings

The requirements of byelaw 2 will be accepted (insofar as individual water fittings and appliances are concerned) as being satisfied in the following circumstances:

    i) **At the time it is connected** in a water service, a water fitting or appliance is capable of complying with any one of the following:

        (a) it lawfully bears a Certification Trade Mark issued by an organisation accredited by the

National Accreditation Council for Certification Bodies, for example the BSI Kitemark (Diagram 7), provided that the British Standard or other specification to which the certification refers is one which is cited in this document or one that conforms with the byelaw requirements; or

(b) it lawfully bears a mark or marks recognised by the Water Industry's Fittings Testing Scheme, for example the Scheme's own mark (see Diagram 8), which should assure purchasers and users, particularly if the manufacturer has been independently assessed and subsequently monitored as conforming to the appropriate parts of British Standard BS 5750 Quality Systems, that it is a replica of a water fitting or appliance which has been accepted as complying with the requirements of byelaws; or

Diagram 7 - BSI Kitemark   **Diagram 8 - Fittings Testing Scheme Mark**

(c) it lawfully bears a mark or marks giving assurance that it is a replica of a water fitting or water appliance which complies with a relevant European Community Directive; or

lawfully bears a mark or marks indicating (in which case the purchaser or user should assure themselves that adequate quality assurance facilities exist with the supplier) that it is a replica of a water fitting or appliance which complies with a relevant European Community Directive; or

(d) it is proven to be capable of passing the appropriate tests set out in the Water Industry's UK Water Fittings Byelaws Scheme's Test and Acceptance Criteria Information and Guidance Note (IGN) 5-50-01 ISSN 0267-0313. Attention is drawn to entries in the Directory and also to various quality assurance schemes available, for example the BSI Registered Firms Scheme, for purchasers and users to obtain continual assurance of conformity to agreed quality standards.

ii) At all times **subsequent to its connection** a water fitting or appliance remains capable of complying with one of the following:

(a) It lawfully bears a mark identified in (i)(a), (b) or (c) and it:

i) has not become altered, modified or changed in any material respect since the time of its first or new connection and

ii) has not become damaged or worn; or

(b) It remains capable of passing the appropriate tests of the Water Industry's UK Water Fittings Byelaws Scheme (see (d) above).

GUIDANCE BYELAWS 2 AND 3

The person connecting, altering or disconnecting water fittings

"Subject to any check or monitoring inspections carried out by the undertakers, work completed by an installer participating in a certification scheme acceptable to the local water undertaking, will be considered as satisfying the requirements of byelaw 2(a) and 3(a). The requirements for such a scheme are set out in IGN 5-01-04. Further advice on plumbers and plumbing contractors may be obtained from the local water undertaking."

14

GUIDANCE BYELAW 3b

Use of faulty water fittings and appliances

The application of byelaw 3(b) depends upon the circumstances which prevail in each case. Leakage may be detected by the undertakers' byelaw or waste control inspectors and a notice served to rectify matters. Similarly, a notice might be issued where water quality has become impaired or is likely to be affected e.g. as a consequence of the use of temporary attachments (such as hoses or anti-splash tubes) or the absence of suitable covers to storage cisterns.

Examples of matters to which these byelaws refer and which must be corrected without delay are:

(a) Dripping taps: waste of water.

(b) Faulty float-operated valves or floats: waste or increased risk of contamination.

(c) Uncovered cisterns or cisterns with unsatisfactory covers: contamination risk.

(d) Leaking pipes and cisterns: waste.

(e) Supply pipes found with their outlets submerged or likely to be submerged in a potentially contaminated liquid: contamination risk.

BYELAW 4

SAVINGS FOR FITTINGS LAWFULLY USED AND FITTINGS USED FOR TEMPORARY NON-DOMESTIC PURPOSES

BYELAW 4. Subject to any express provisions to the contrary, these byelaws shall not -

(a) require any person to remove, replace, alter, disconnect or cease to use any water fitting lawfully installed, or lawfully used, or capable of being used, before these byelaws came into operation; or

15

GUIDANCE BYELAW 4

The requirements of byelaw 4(b)(iii) will be accepted as being satisfied
where the measures to ensure water cannot return into the supply pipe
accord with Part III of the byelaws.   Especially see the comment to
byelaw 25.

16

PART II

PREVENTION OF CONTAMINATION OF WATER FROM CONTACT WITH UNSUITABLE MATERIALS OR SUBSTANCES

CONTENTS

PART II - PREVENTION OF CONTAMINATION OF WATER FROM CONTACT WITH UNSUITABLE MATERIALS OR SUBSTANCES

BYELAW 5

PROHIBITION OF INSTALLATION OF PIPES IN CONTACT WITH CONTAMINATING MATERIALS

> BYELAW 5. No supply pipe, distributing pipe or other water fitting shall be laid or installed in or on, or pass into or through any foul soil, refuse or refuse chute, ash pit, sewer, drain, cesspool, or any manhole connected with any such sewer, drain or cesspool.

GUIDANCE BYELAW 5

This prohibition is made regardless of any protection to the pipe. Diagrams 9 and 10 illustrate correct and incorrect methods of laying pipes in the vicinity of manholes.

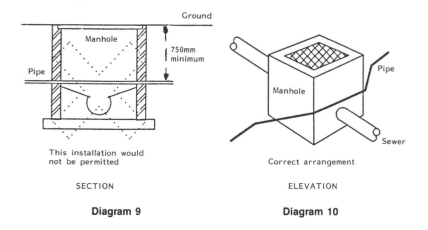

This installation would not be permitted	Correct arrangement
SECTION	ELEVATION
Diagram 9	**Diagram 10**

19

BYELAW 6

PERMEATION AND DETERIORATION OF PIPES

BYELAW 6. No supply pipe, distributing pipe or other water fitting made of any material which is susceptible to -

    (a) permeation by any gas which causes, or is likely to cause, or

    (b) deterioration by contact with any substance which causes, or is likely to cause;

contamination of the water in that pipe or fitting, shall be laid or installed in a place or position where such permeation or deterioration occurs, or is reasonably likely to occur.

GUIDANCE BYELAW 6

Because some plastics pipes can be permeated by gas, care should be taken regarding the circumstances in which they are used. Diagram 11 is a recommendation by the National Joint Utilities Group Report 6 for service entries for new dwellings. It provides in effect a space of about 350mm between gas and water services.

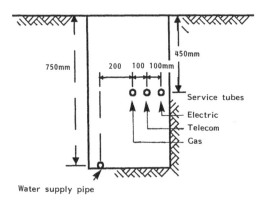

Diagram 11

BYELAW 7

MATERIALS IN CONTACT WITH WATER

> **BYELAW 7**   (1) No material or substance which causes, or is likely to cause, contamination of water shall be used in the construction or installation of any pipe or water fitting which conveys or receives water supplied for domestic purposes.
>
>   (2) Paragraph (1) shall not apply to -
>
>   (a) any hosepipe used in connection with a clothes washing machine or dishwasher, or for watering a garden, or washing a vehicle kept for private use, where the pipe or other fitting to which that hosepipe is, or may be, connected incorporates a check valve or some other no less suitable device to prevent the backflow or backsiphonage of water through that hosepipe; or
>
>   (b) any flushing cistern; or
>
>   (c) any feed cistern connected to a primary circuit; or
>
>   (d) any closed circuit; or
>
>   (e) any warning pipe.

GUIDANCE BYELAW 7(1)

General standards of acceptance

Substances leached from materials of construction of pipes, cisterns or other water fittings in contact with water must not adversely affect water quality at the kitchen tap or drawn from cisterns storing water for drinking or culinary purposes. See also guidance to byelaw 96 concerning water softeners.

The requirements of byelaw 7(1) would be accepted as being satisfied if sampling shows compliance with either:

        (a) the maximum values recommended by the World Health Organisation Guidelines for

Drinking Water Quality: Volume 1 "Recommendations" (WHO Geneva 1984); or

(b) EC Directive of 15 July 1980 - Quality of water intended for human consumption (OJL 229 pp11 to 29) generally obtainable from public libraries.

Requirements for testing non-metallic materials are set out in BS 6920.

Acceptable materials

Products manufactured for installation and use in the United Kingdom which are listed in the Directory, are accepted as satisfying the requirements of this byelaw as are materials listed by the Department of the Environment Committee on Chemicals and Materials of Construction for use in Public Water Supply and Swimming Pools which should be considered free from toxic hazard. The latter list is available from the Technical Secretary of that Committee at the Department of the Environment, Water Technical Division, Romney House, 43 Marsham Street, London SW1P 3PY. Note that these documents do not cover metallic materials.

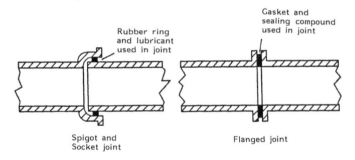

Diagram 12 - Examples of materials covered by byelaw 7(1)

Pending the publication of a British Standard for lead-free solders suitable for use in jointing copper pipes in new domestic water installations, the use of solders not containing lead, e.g. tin/silver, will be acceptable.

In any building or part of a building which has already been connected to the mains in which water is used for domestic purposes, if pipework upstream of any draw-off tap is found by an Inspector to have been jointed by solder containing lead, the owner/occupier of the premises may be advised about the potential risk. This will be without prejudice to other courses of action open to the undertakers.

GUIDANCE BYELAW 7(2)(a)

This proviso to the byelaw allows hoses not complying with byelaw 7(1) to be connected to a water service in certain cases. An example is given in diagram 13. Hoses listed in the Directory do not require such a check valve.

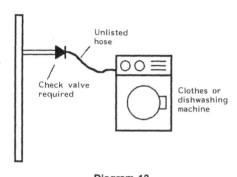

Diagram 13

GUIDANCE BYELAW 7(2)(b) to (e)

Proviso 7(2)(b) exempts flushing cisterns where it is common practice to use additives, etc. The remaining provisos exempt cisterns and pipes connected to closed circuits (see diagram 14) where inhibitors can be used without risk of contamination of drinking water providing adequate precautions are taken to prevent backflow as set out in Part III.

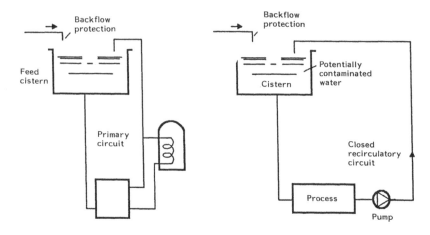

Diagram 14 - Exemptions under byelaws 7(2)(c) and (d)

BYELAWS 8 AND 9

PROHIBITION OF COAL TAR AND LEAD

BYELAW 8. No pipe, pipe fitting or storage cistern shall be internally lined or coated with coal tar, or any substance which includes coal tar.

BYELAW 9. No lead pipe or other water fitting, or storage cistern made from, or internally lined with, lead shall be installed (whether or not by way of repair or replacement of a similar pipe or cistern) after the commencement of these byelaws.

GUIDANCE BYELAWS 8 AND 9

Leachates from coal tar and lead have been identified as potential health hazards. Reference should be made to the Directory for a list of alternative materials acceptable for these purposes. See also guidance on byelaw 7(1) above.

24

BYELAW 10

PROHIBITION OF COPPER IN REPAIRS TO LEAD PIPE

BYELAW 10. No copper pipe shall be connected to or incorporated with any lead pipe (whether or not by way of repair or replacement) unless suitable means are employed to prevent, so far as is reasonably practicable, corrosion through galvanic action.

GUIDANCE BYELAW 10

It is recommended that in any event copper pipe is not used for repairing lead pipes. Fittings are available and are listed in the Directory for connecting plastics pipe into a lead pipe. See also guidance to byelaw 60.

Warning: Care should be taken when cutting out part of a metallic service pipe. This may affect the earthing of an electrical system in fault conditions and give an electric shock. This is not a byelaw matter.

PART III

PREVENTION OF CONTAMINATION OF WATER BY BACKSIPHONAGE, BACKFLOW OR CROSS CONNECTION

CONTENTS

27

PART III - PREVENTION OF CONTAMINATION OF WATER BY BACKSIPHONAGE, BACKFLOW OR CROSS CONNECTION

INTRODUCTION

The byelaws included in this section are concerned with the protection of all mains supplied water services from the risk of contamination arising from backflow.   The section commences with definitions of backflow prevention devices.   Only a few devices are described in the byelaws or in this Guide.   As others are developed, tested and accepted they will be listed under appropriate "risk" categories in the Directory.

The first principle of protection is that mains supplied water is kept separate from any other water e.g.   that supplied from a private source or non-potable water.   This applies equally to water supplied from the mains but which, even if the quality has not been significantly impaired, has been drawn off.   This principle is dealt with in **Byelaw 12.**   Note that water taken from the supply pipe into a cistern, other than a cistern that complies with byelaw 30(2) has been drawn off and used.

The second principle is that there should be no cross connections within an installation, e.g.   from a supply pipe to a distributing pipe, thus ensuring that potentially contaminated water in a distributing system cannot reach the supply pipe.   Nor should a pump or other device be connected so as to cause water to flow back into the supply pipe.   See **Byelaws 13 and 14**

The third principle is that protection against backflow should be afforded at every point of use be it a draw-off tap, a flushing cistern or a washing machine.   This protection is secured by the use of a device designed to match the particular risk arising.   **Byelaws 16 to 24** set out requirements for such protection in the specific cases of draw-off taps, hose connections, bidets, clothes and dishwashing machines, water softeners and cisterns.

Byelaw 25 is the general requirement for such protection in all other cases.

Protection at point of use is based on three categories of risk detailed in **Byelaw 25** i.e.   in 25(1)(a), 25(1)(b) and 25(2) respectively and for which three categories of protection are appropriate.   These categories are derived from the report of the DoE Committee on Backsiphonage in Water Installations (HMSO 1974).

In the guidance to byelaw 25, schedules A to C categorise risks for a number of the more common situations. Further information is given in BS 6700.

The fourth principle is that in addition to the basic protection provided by prevention of cross connection and by protection at points of use further protection is required by **Byelaw 26** except in the case of separately occupied premises connected to the mains by a separate supply pipe. Such additional protection would reduce the risk of contaminated water being drawn, for example, from one flat to another flat at a lower level. It is termed **secondary backflow protection**.

BYELAW 11

DEFINITIONS

BYELAW 11. In this part of the byelaws -

BACKFLOW PREVENTION DEVICE means either a type A or type B air gap, a check valve, a double check valve assembly, a combination of check valve and vacuum breaker, a pipe interrupter, or some other water fitting or arrangement of water fittings designed to prevent the backflow or backsiphonage of water;

CHECK VALVE means a mechanical device which -

    (a) by means of a resilient elastic seal or seals permits water to flow in one direction only and is closed when there is no flow; and

    (b) (i) is resistant to corrosion, and

        (ii) is immune from, or resistant to, dezincification, and

        (iii) will continue to operate in a continuous water temperature not exceeding 65°C, and

        (iv) when shut will prevent the passage of water from inlet to outlet where the water pressure at the valve inlet does not exceed 10 mbar;

CRITICAL WATER LEVEL in relation to a type B air gap means the steady water level in a cistern, vessel or other water fitting when there is a maximum inflow of water and all outlets, except any overflow, are closed;

DOUBLE CHECK VALVE ASSEMBLY means a mechanical device comprising two check valves with a test cock between them;

PIPE INTERRUPTER means a non-mechanical device without any moving, flexible or elastic parts which -

(a) in the event of any vacuum in a pipe in which it is installed will admit air into it to prevent the backflow of water; and

(b) (i) is resistant to water corrosion and is immune from, or resistant to, dezincification, and

(ii) has an unobstructed air inlet-aperture, or apertures which, when a vacuum occurs on the inlet side, produces a corresponding vacuum on the outlet side not exceeding 5 mbar below atmosphere;

TYPE A AIR GAP occurs if there is an arrangement of water fittings whereby -

(a) water is discharged into a cistern, vessel or other fitting which has at all times an unrestricted overflow to the atmosphere; and

(b) the pipe discharging into that cistern, vessel or other water fitting is not obstructed; and

(c) water is discharged downwards into the cistern, vessel or other fitting at not more than 15° from vertical; and

(d) the vertical distance from the spill-over level of the unrestricted overflow of that cistern, vessel or other fitting to the point above that spill-over level which is the lowest point of any pipe or fitting which discharges into that cistern or vessel or fitting is not less than the figure mentioned in the Table below in relation to a pipe of the appropriate bore.

TABLE

1. Bore of pipe or outlet	2. Vertical distance of point of outlet above spill-over level
1. not exceeding 14mm	20mm
2. exceeding 14mm but not exceeding 21mm	25mm
3. exceeding 21mm but not exceeding 41mm	70mm
4. exceeding 41mm	twice the bore of the outlet.

TYPE B AIR GAP occurs when water is discharged into a cistern, vessel or other water fitting which is open at all times to the atmosphere, and the vertical distance between the lowest point of discharge into that cistern, vessel or water fitting and its critical water level is either -

(i) sufficient to ensure that, if there were a vacuum in that discharge pipe or fitting, no water in the cistern, vessel or water fitting would be siphoned back into that pipe or fitting, or

(ii) not less than the figure mentioned in the preceding Table in relation to a pipe of the appropriate bore.

VACUUM BREAKER means a mechanical device with an air inlet which is closed when water flows past it at or above atmospheric pressure but which opens to admit air if there is a vacuum in the pipe and closes so as to be watertight when the flow of water is resumed at normal pressure.

GUIDANCE BYELAW 11

The backflow prevention devices mentioned in the byelaws are limited in number to those considered acceptable for the time being. As others are developed, tested and found acceptable they will be listed under appropriate "risk" categories in the Directory. The following are illustrative of the principles of operation of the backflow prevention devices mentioned.

1. Type A air gap (diagram 15)

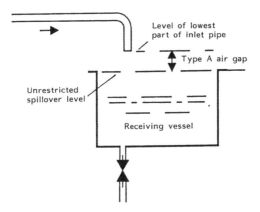

Diagram 15 - Example of Type "A" air gap

Type A air gaps would be accepted as satisfying byelaw 11 where any air gap complies with BS 6281: Part 1. "Unrestricted" in this byelaw means that no object is closer than 3 times the bore of the feed pipe to the feed pipe or to the vertical projection of the feed pipe between that pipe and the top edge of the receiving vessel or other water fitting.

2. Type B air gap (diagram 16)

The requirements of byelaw 11 would be accepted as being satisfied where any Type B air gap complies with BS 6281: Part 2.

In the case of a cistern fitted with a float-operated valve on the feed pipe the critical water level referred to in byelaw 11 should be established as the highest steady water level which occurs when the float is either punctured or removed at a time when the valve is conveying the maximum flow of water.

In a case where an appliance has a built in Type B air gap and is to be tested in accordance with BS 6280 the maximum rate of flow should be taken as that resulting from the lesser of either an inlet pressure of 6 bar or a flow equivalent to a velocity of 2m/sec in the feed pipe.

The highest water level shown in diagram 16 is the critical water level defined in byelaw 11 and in the British Standard.

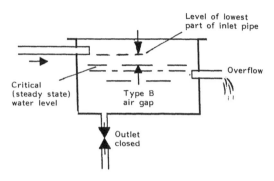

Diagram 16 - Type B air gap

3. Check valve (diagram 17)

The requirements of byelaw 11 would be accepted as being satisfied where a check valve complies with BS 6282: Part 1.

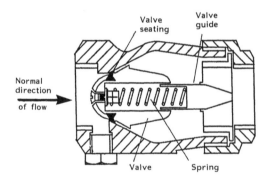

Diagram 17 - Check valve

4. Double check valve assembly

A double check valve assembly would be accepted as satisfying byelaw 11 if it consists of an assembly of two check valves to BS 6282: Part 1 with an intervening draining tap to BS 2879.

5. Pipe interrupter (diagram 18)

The requirements of byelaw 11 would be accepted as being satisfied if a pipe interrupter complies with BS 6281: Part 3. A pipe interrupter has no moving parts.

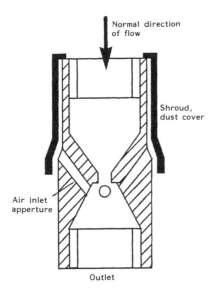

Diagram 18 - Example of pipe interrupter

6. Vacuum breaker (diagram 19)

The Report of the Committee on Backsiphonage (HMSO 1974) referred to "anti-vacuum valves" which are now described in the byelaws as vacuum breakers.

There are two basic types of in-line vacuum breaker. The "atmospheric" type which is dealt with in BS 6282: Part 3 needs to be regularly opened and shut to keep it operable and for this reason needs to drain down and open to admit air after use; thus there should be no control valve downstream. The "pressure" type has a spring loaded air inlet biased to the open position and a spring loaded check element biased to a closed position and is said to operate satisfactorily even after having been under continuous pressure for long periods. Only the former type is recognised in this guide and at the present time there is no British Standard for the "pressure" type of vacuum breaker which is under investigation at the time of publication. Both types must be installed at least 150mm above the highest level of any contaminated water because neither will give protection under back pressure backflow.

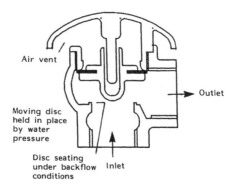

Diagram 19 - Example of an "atmospheric" type vacuum breaker

Note that the above BS states "the provision of conducting any water spilled from the device is optional". This is because the air inlet ports can provide a means for the escape of water in fault conditions.

7. Combination of a check valve and a vacuum breaker

A combination of a check valve to BS 6282: Part 1 and an in-line vacuum breaker satisfying BS 6282: Part 3 would be accepted as an alternative to a double check valve assembly providing that:

(a) the vacuum breaker element is situated downstream of the check valve (see comment under vacuum breaker),

(b) there is no control valve downstream of the vacuum breaker,

(c) the vacuum breaker is installed on a prescribed upstand, see diagram 20.

Note also the advice in Section 6 given on dealing with escaping water.

8. Upstands on supply and distributing pipes (diagram 20)

Although not mentioned in byelaw 11, upstands are an implicit requirement to provide additional protection in many cases, for example when associated with a vacuum breaker.

The dimension of an upstand is determined by the degree of vacuum likely to be encountered at the point of connection with the supply or distributing pipe. It can be limited by the operation of a vacuum breaker or a vent pipe which admits air.

The effect of the upstand would be to prevent contaminated liquid being drawn from an appliance into the supply or distributing pipe should protection at point of use fail.

See also interpretation of vent pipe in Part I.

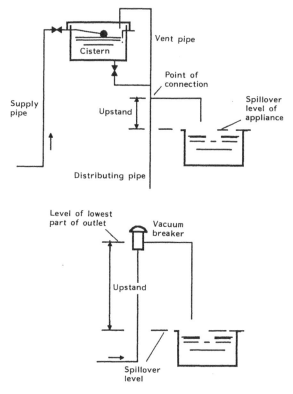

Diagram 20 - Upstands

BYELAW 12

TAKING OF SUPPLIES

BYELAW 12. (1) No supply or distributing pipe which conveys, or cistern which receives, water supplied for domestic purposes shall be connected so that it can convey or receive water supplied for non-domestic purposes.

(2) Paragraph (1) shall not apply to a cistern, or to any pipe conveying water from such a cistern to a point of use if water is discharged into that cistern through a Type A air gap.

GUIDANCE BYELAW 12

GENERAL

Byelaw 12 protects the supply pipe from water from any other source including water supplied by the undertaker for industrial purposes, water from any private source, or water supplied for domestic purposes that has been drawn off. The byelaw is illustrated in diagram 21. In no event must a connection be made between a supply pipe and a pipe carrying water supplied for non-domestic purposes but byelaw 12(2) permits the latter to be connected to a cistern or a distributing pipe if the supply pipe discharges to the cistern through a Type A air gap. In this event water in the cistern and in the distributing pipe can no longer be considered as water supplied for domestic purposes.

Where a supply is made direct from the undertakers' mains solely for non-domestic purposes, the undertaker will require that the mains are protected from any contamination which may occur in the installation. The precautions relevant to supplies made for domestic purposes can be taken as a guide to what may be required.

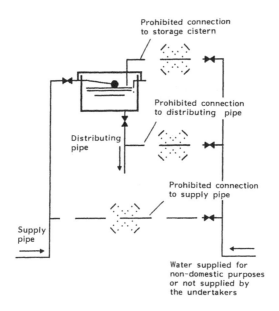

Prohibited connection
to storage cistern

Prohibited connection
to distributing pipe

Distributing
pipe

Prohibited connection
to supply pipe

Supply
pipe

Water supplied for
non-domestic purposes
or not supplied by
the undertakers

Diagram 21 - Connections prohibited under byelaw 12

Sections 1 to 5 below give examples of ways in which supplies may be taken meeting the requirements of byelaw 12. Reference is made in some of the diagrams to BS 1212: Part 2 or Part 3 float operated valves as an alternative to a Type B air gap. Because such valves are capable of withstanding a suitable vacuum test and are of the reducing flow type, they afford some protection against backflow and the installation requirements do not need to be as stringent as for a Type B air gap. As an alternative to a Type B air gap they would only be accepted if flow from the cistern were by gravity at all times and where the centre line of the valve was not lower than the centre line of the warning pipe. See also under byelaw 24.

1. Domestic supplies

(a) Taking of domestic supplies in a dwelling (diagram 22).

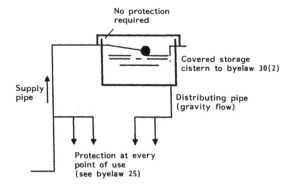

Diagram 22

(b) Taking of domestic supplies within industrial and commercial premises, hospitals, medical establishments, farms, etc. where non-domestic supplies are also taken (diagram 23).

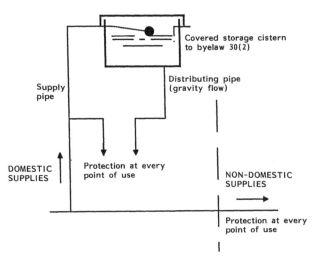

Diagram 23

Should non-domestic supplies be taken through a cistern complying with byelaw 24(1)(a) the requirement for protection at every point of use beyond that cistern would not apply.

2.   Non-domestic supplies within industrial and commercial premises, hospitals, medical establishments, etc.

(a) Separation of mains' supplies from other supplies (diagram 24).

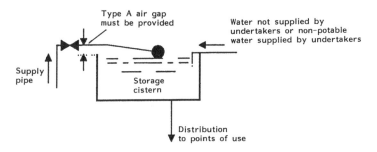

Diagram 24

(b) Separation of water in a supply pipe from water that has been used (diagram 25).

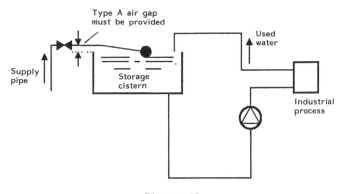

Diagram 25

41

(c) Supply taken into a storage cistern with gravity flow from it at all times (diagram 26). See also guidance under byelaw 24(2). Should the flow from the cistern be by means other than by gravity to points of use the inlet feed pipe to the cistern must be arranged to provide a Type A air gap.

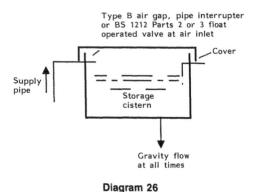

Type B air gap, pipe interrupter or BS 1212 Parts 2 or 3 float operated valve at air inlet

Cover

Supply pipe

Storage cistern

Gravity flow at all times

Diagram 26

NOTE The type of protection illustrated in diagram 26 might be required adjacent to risks in particularly hazardous situations such as a pathology laboratory. In such an instance the supply must be taken from storage. No other areas or drinking water fittings are to be supplied from that storage which, if in the laboratory, must be covered and have appropriate protection at the inlet.

(d) Supply taken directly to a fixed or mobile appliance in industrial, commercial, etc. premises (diagram 27).

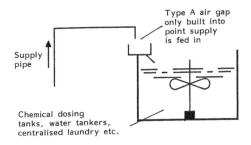

Type A air gap
only built into
point supply
is fed in

Supply
pipe

Chemical dosing
tanks, water tankers,
centralised laundry etc.

Diagram 27

3. Domestic supplies taken by ships, boats, etc.

(a) Supply taken direct from mains to ship or boat (diagram 28).

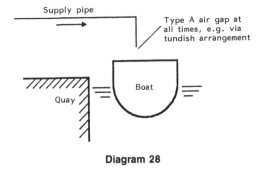

Supply pipe

Type A air gap at
all times, e.g. via
tundish arrangement

Quay

Boat

Diagram 28

NOTE   Drinking and non-potable watering points must be segregated and readily distinguishable from each other by suitable marking of the supply pipe (see comment to byelaw 27).

(b) Supply taken by gravity from storage on shore (diagram 29).   See note in 3(a) above.

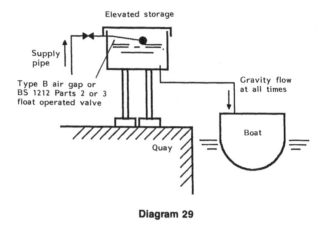

Diagram 29

(c) Supply pumped from storage on quayside (diagram 30). See note in 3(a) above.

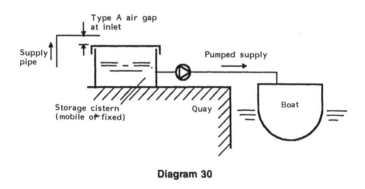

Diagram 30

Diagram 30

(d) Small supplies taken from a stand pipe at a marine, etc. through a hose not exceeding 22mm nominal diameter (diagram 31). See notes in 3(a) above.

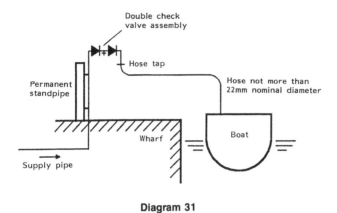

Diagram 31

NOTE Mechanical devices should be protected against frost. For alternatives to a double check valve assembly see byelaw 11 or consult the Directory.

4. Non-domestic supplies taken in agriculture, horticulture, etc.

(a) Supplies taken from storage (diagram 32).

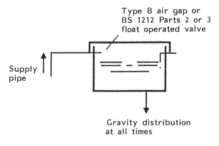

Diagram 32

NOTE Should water from any other source be admitted to this storage or the cistern is uncovered then a Type A air gap would be required at the inlet.

(b) Supplies taken from storage by pump or to plant in which pressure could be increased (diagram 33).

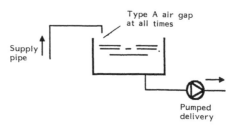

Diagram 33

(c) Supplies taken direct from supply pipe to fixed or mobile plant (diagram 34).

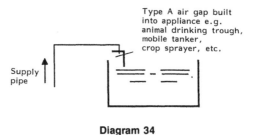

Diagram 34

NOTE In the case of animal drinking troughs the inlet must be shrouded to prevent access by animals. Correctly installed drinking troughs to BS 3445: Fixed agricultural water troughs and water fittings, will be accepted as satisfying this requirement.

(d) Supplies taken from the supply pipe by hose.

(i) A Type A air gap must be maintained at all times at the outlet end of the hose: i.e. the end of the hose must be fixed to ensure an air gap above any used water.

46

(ii) If no Type A air gap can be provided at the outlet end of the hose the supply must be taken via storage complying with (a) above: i.e. a hand held hose cannot be deemed to ensure compliance at all times.

(iii) These provisions need not apply to any hose tap connection which is accepted by the undertakers as not posing an unacceptable risk of contamination of the mains and where every such tap is permanently fitted with a double check valve assembly or some other no less suitable device.

5. Water taken for fire sprinkler systems

Attention is drawn to byelaw 28 prohibiting the use of pipes installed for fire fighting purposes from being used for other purposes.

(a) Wet system directly connected under mains pressure (diagram 35).

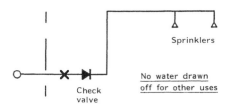

Diagram 35

(b) Wet system directly connected under mains pressure (diagram 36).

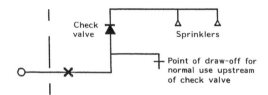

Diagram 36

(c) Water taken to elevated storage (diagram 37). No other water is to be supplied to that storage and the water quality must be preserved by sufficient water being taken for other non-domestic purposes.

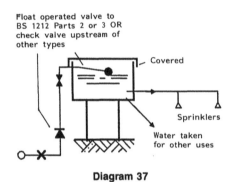

Diagram 37

(d) Water taken to elevated storage (diagram 38). No other water to be supplied to that storage. In this case a Type A air gap at the inlet would be recommended and in most cases this should be capable of being installed.

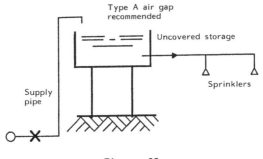

Diagram 38

(e) Water pumped from storage (diagram 39).

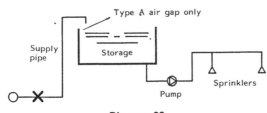

Diagram 39

(f) Water taken to storage supplemented with supply from another source (diagram 40).

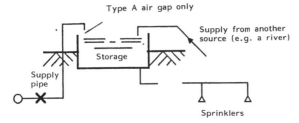

Diagram 40

49

(g) Improvement of protection of an existing installation which no longer complies with the byelaws. Correct existing arrangement as shown on diagram 41.

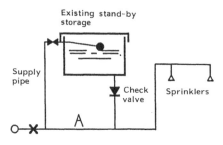

Diagram 41

NOTE   This only applies to existing installations and is not accepted for new ones.   It is in no way intended for application if the connection "A" can be removed without reducing the effectiveness of the fire sprinkler system.   For reduced pressure principle backflow preventer see reference in guidance to byelaw 25 (devices under investigation).

BYELAW 13

PREVENTION OF CROSS CONNECTIONS

BYELAW 13.   (1) No pump or other apparatus shall be connected in or to a supply pipe for the purpose of increasing the pressure in, or rate of flow from -

(a) a service pipe; or

(b) any water fitting connected on or to a service pipe;

except with the prior written consent of the undertakers.

(2) No supply pipe or pump delivery pipe drawing water from a supply pipe shall

convey, or be connected so that it can
convey, water from -

(a) a distributing pipe; or

(b) a storage or flushing cistern; or

(c) a pump delivery pipe drawing water
    from a distributing pipe or cistern; or

(d) a pipe or vessel pressurised by
    compressed air or gas; or

(e) a source other than from the
    undertakers' mains.

GUIDANCE BYELAW 13(1)

Because pumps can increase pressure above that in the supply pipe
and in the mains there is an enhanced risk of a discharge of possibly
contaminated water back into the supply pipe.

The written consent of the undertakers would be deemed to have been
given in the case of pumps incapable of drawing more than 10 litres
per minute installed in drink vending and dispensing machines and
domestic water softeners in other respects fully complying with the
byelaws.

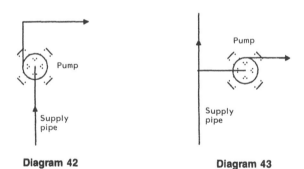

Diagram 42   **Diagram 43**

Examples of prohibited pump connections

51

GUIDANCE BYELAW 13(2)

This byelaw is designed to prevent cross-connections within an installation which might contaminate a supply pipe. The following are examples of prohibited connections.

(a) No supply pipe shall be connected to a distributing pipe (diagram 44) Byelaw 13(2)(a).

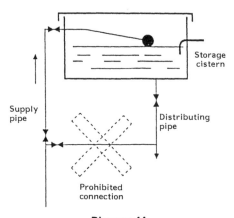

Diagram 44

(b) No supply pipe shall be so connected that it can draw water from a storage cistern (diagram 45) Byelaw 13(2)(b).

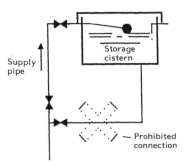

Diagram 45

(c) No pump shall be able to convey water into a supply pipe from a distributing pipe or a storage cistern (diagram 46) Byelaw 13(2)(c).

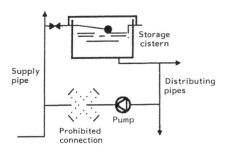

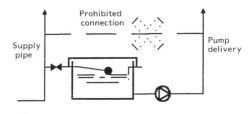

Diagram 46

(d) No vessel containing compressed air or gas shall be able to convey water into a supply pipe. Note that this does not apply to surge suppressors or expansion vessels (diagram 47) Byelaw 13(2)(d).

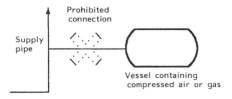

Diagram 47

BYELAW 14

CONNECTIONS TO CLOSED CIRCUITS

BYELAW 14. (1) No closed circuit shall be connected to a supply pipe. (2) Paragraph (1) shall not apply to a temporary connection provided - (a) the connection is made through a double check valve assembly or some other no less effective device which is permanently connected to that circuit; and (b) the temporary connection is removed after use.

GUIDANCE BYELAW 14

This byelaw is concerned with the filling of primary heating circuits. Because water in primary circuits of heating systems often contains additives and the water can be heavily contaminated, direct connections to the supply pipe are prohibited except in circumstances detailed below. In addition there must be no cross-connection between a primary (water heater) circuit and a secondary circuit from which water is drawn off for use (see Byelaw 15).

Make up water to the primary circuit can only be supplied from a separate feed cistern although it will be permitted to fill or top up a primary circuit from a supply pipe provided it meets the proviso in byelaw 14(2). Note that care should be taken that heat does not impair the operation of the device.

A double check valve assembly would be accepted until alternative devices are available and are listed for the purpose in the Directory. The arrangement is illustrated in diagram 48. For connection to the supply pipe see also byelaw 18.

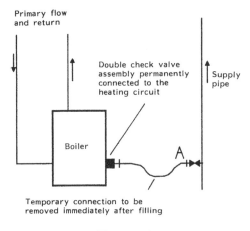

Primary flow
and return

Double check valve
assembly permanently
connected to the
heating circuit

Supply
pipe

Boiler

A

Temporary connection to be
removed immediately after filling

Diagram 48

NOTE If the hose connection to the supply pipe is to be used for any other purpose, protection prescribed in byelaw 18 will be required at A in addition to that connected to the heating circuit.

BYELAW 15

PREVENTION OF CONNECTION BETWEEN CISTERN FED PRIMARY CIRCUITS AND SECONDARY SYSTEMS

BYELAW 15. No pipe forming part of a cistern fed vented primary circuit shall be connected to any pipe forming part of a secondary system.

GUIDANCE BYELAW 15

Diagram 49 illustrates the application of this byelaw. But see byelaw 37 with regard to the installation of single feed indirect cylinders.

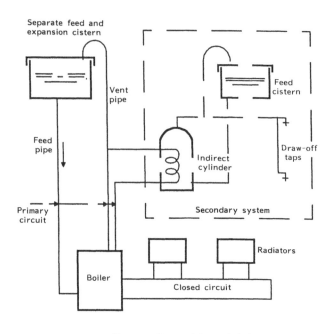

Separate feed and
expansion cistern

Vent pipe

Feed cistern

Feed pipe

Indirect cylinder

Draw-off taps

Primary circuit

Secondary system

Radiators

Boiler

Closed circuit

No connection must be made between

the primary circuit —————— and the

secondary system ——   ——

Diagram 49

BYELAW 16

DRAW-OFF TAPS TO BATHS, SINKS, ETC.

BYELAW 16.	(1) Subject to paragraph (3), every draw-off tap or similar fixed fitting (other than a shower hosepipe) installed to discharge water into a bath, sink, washbasin or similar fixed appliance (other than bidet) shall -
	(a) incorporate a double check valve assembly; or

56

(b) have fitted as close as is practicable to the point of draw-off or use, some other no less effective device to prevent the backflow or backsiphonage of water; or

(c) comply with paragraph (2).

(2) For the purpose of paragraph (1)(c) the vertical distance from the spill-over level of that bath, sink or other appliance to the point above that spill-over level which is the lowest point of any draw-off tap or other fixed fitting which discharges into that bath, sink or other appliance is not less than the figure mentioned in the table below in relation to a tap of the appropriate size.

TABLE

1. Size of tap, or combination fitting	2. Vertical distance of point of outlet above spill-over level
1. Not exceeding $1/2$ inch	20mm
2. exceeding $1/2$ inch but not exceeding $3/4$ inch	25mm
3. exceeding $3/4$ inch	70mm

(3) Paragraph (1) shall not apply to any draw-off tap or other water fitting where -

(a) that tap or fitting draws water by gravity only from a cistern, cylinder or tank having a vent pipe open at all times to the atmosphere; and

(b) the vertical distance between the point at which the pipe supplying water to that tap or other fitting connects to the cistern, cylinder or tank and the spill-over level of the relevant bath, sink or other appliance is not less than 25mm; and

(c) the pipe supplying water to that tap or other fitting does not supply water to any other tap or fitting (other than a draining tap) at a lower level.

Subject to site verification of the air gap with the table in byelaw 16, the requirements of byelaw 16(2) would be accepted as being satisfied in respect of a combination of a draw-off tap and appliance complying with the appropriate requirements of any of the standards mentioned below. This is illustrated in diagram 50. In due course BS 1010 taps (the standard is under amendment) are expected to qualify under this byelaw.

BS 5412:   Specification for the performance of draw-off taps with metal bodies for water services.

BS 5413:   Specification for the performance of draw-off taps with plastics bodies for water services.

FITTED TO:

BS 1244:   Metal sinks for domestic purposes Part 2: Stainless steel sink tops.

BS 1390:   Sheet steel baths.

BS 1189:   Baths made from porcelain enamelled cast iron.

BS 4305:   Baths for domestic purposes made from cast acrylic sheet.

BS 1188:   Ceramic washbasins and pedestals.

BS 5506:   Washbasins Part 3: Washbasins (one or three tap holes).

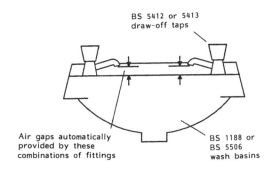

BS 5412 or 5413
draw-off taps

Air gaps automatically
provided by these
combinations of fittings

BS 1188 or
BS 5506
wash basins

Diagram 50 - Compliance with byelaw 16(3)

If the tap arrangement is such that an air gap to the table as determined by site measurement could not be achieved then a double check valve assembly would be accepted fitted as close as practicable to the draw-off point. This is illustrated in diagram 51.

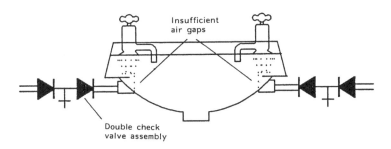

Insufficient
air gaps

Double check
valve assembly

Diagram 51

In any case where both hot and cold water is mixed before discharge and an air gap cannot be provided, if the installer proposes to connect single check valves on the hot and cold feed pipes, a third check valve connected at the mixed water outlet will be accepted as meeting this requirement. This arrangement would protect against contamination from the sink or basin as well as the mixing of hot and cold water in the feed pipe.

The proviso in byelaw 16(3)(c) exempts certain cases where a tap is the lowest draw-off point as shown in diagram 52.

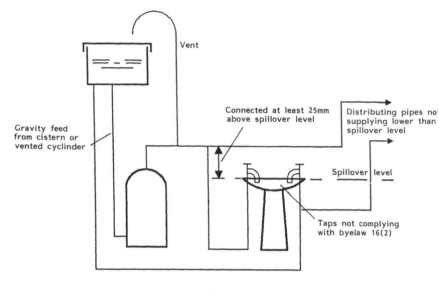

Diagram 52

BYELAW 17

SHOWER HOSE CONNECTIONS

BYELAW 17.

(1) Subject to paragraph (2), every draw-off tap or other water fitting which incorporates a shower hosepipe (whether or not operated by a manual or automatic diverter) installed to discharge water into a bath, shower tray, sink, washbasin or similar fixed appliance (other than a bidet) shall -

(a) incorporate a double check valve assembly; or

(b) incorporate a check valve and a vacuum breaker; or

(c) have fitted as close as is practicable to the point of drawoff or use some other no less effective device to prevent the backsiphonage or backflow of water.

(2) Paragraph (1) shall not apply to any draw-off tap or other fitting where -

(a) the tap or other fitting is installed in accordance with paragraph (3) of byelaw 16; or

(b) the shower head of any shower hosepipe is constrained by a fixed or sliding attachment so that it can only discharge water at a point not less than 25mm above the spill-over level of the relevant bath, shower tray or other fixed appliance; or

(c) the vertical distance between the shower head of any unconstrained shower hosepipe and the spill-over level of the relevant bath, shower tray or other fixed appliance is not less than the figure mentioned in the Table in byelaw 16(2) in relation to a tap or fitting of the appropriate size.

GUIDANCE BYELAW 17

Some examples of arrangements which will be accepted as complying with this byelaw are given below:

Shower hose capable of being lowered below a spillover level

(a) Unvented water heater (diagram 53).

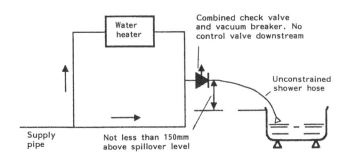

Diagram 53

61

NOTE The vacuum breaker must be high enough to drain down through the hose after every use if the breaker is of the atmospheric type.

(b) Shower complying with byelaw 16(3) (diagram 54). Proviso in byelaw 17(2)(a)

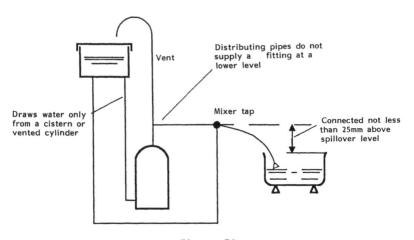

Diagram 54

Shower hose incapable of being lowered below a spillover level

Diagram 55 illustrates byelaw 17(2)(c).

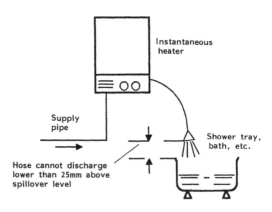

Diagram 55

BYELAW 18

HOSE CONNECTIONS

BYELAW 18. (1) No hosepipe, other than a hosepipe used in accordance with byelaw 7(2) or a shower hosepipe installed in accordance with byelaw 17, shall be connected to a draw-off tap or other similar fitting for use either inside or outside any premises.

(2) Paragraph (1) shall not apply to any hosepipe connected to a draw-off tap or other similar fitting which -

(a) draws water by gravity only from a cistern by means of a pipe which does not supply water to a draw-off tap or similar fitting (other than a draining tap) at a lower level; or

(b) is on domestic premises, or elsewhere with the written consent of the undertakers, and incorporates as close as is practicable to the point of draw-off or use either a double check valve assembly or some other no less effective backflow prevention device.

GUIDANCE BYELAW 18

In domestic premises hoses may be connected to draw-off taps protected by double check valve assemblies or other no less effective devices.   These latter will be listed in the Directory as they become available and are acceptable.   Acceptable arrangements are illustrated in diagram 56; these do not show the thermal insulation which should be arranged as required in Part V.

In other premises hose connections may only be made to draw-off taps, etc. drawing water by gravity from a cistern unless the undertaker gives written consent in a particular case.   This also applies to fire hose reels.

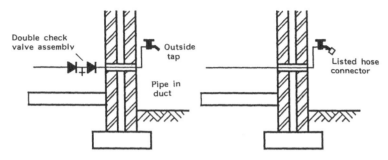

Diagram 56 - Outside connections for hoses in domestic premises only (thermal insulation omitted. See guidance byelaw 49)

BYELAWS 19 TO 21

BIDETS

BYELAW 19. (1) No bidet which is equipped with any type of submersible spray, or any draw-off fitting to which a hand held flexible spray is attached, shall be connected to any supply pipe.

(2) No hand held flexible spray or similar fitting shall be attached to any draw-off fitting on a bidet connected to any supply pipe.

BYELAW 20. Subject to byelaw 19, every bidet connected to a supply pipe -

(a) shall be of the over-rim water feed type; and

(b) shall be installed so that the vertical distance between the outlet point of any draw-off tap or similar fitting and the spillover level of the bidet is not less than the figure mentioned in the Table in byelaw 16(2) in relation to a tap or fitting of the appropriate size.

BYELAW 21. (1) No hot water pipe or water heater which supplies hot water, and no distributing pipe which supplies cold water, to a bidet shall also supply water to any other draw-off tap or similar fitting (except a draining tap) which can discharge water at a point below the spill-over level of the bidet.

64

(2) Paragraph (1) shall not apply -

(a) to a pipe supplying water to a bidet which complies with byelaw 20; or

(b) to a distributing pipe which supplies cold water only to a flushing cistern or to a urinal; or

(c) to a hot water pipe which supplies water only to a bidet and which either -

(i) has a check valve and vent pipe arranged to prevent the backflow or backsiphonage of water from a bidet, or

(ii) is fitted with some other no less effective backflow prevention device.

GUIDANCE BYELAWS 19 TO 21

Byelaw 19 precludes the connection of rim feed or ascending spray type bidets or bidets with hand held spray attachments to any supply pipe and byelaw 20 covers the use of over-rim type feeds, specifying air gap requirements. Byelaw 21(1) lays down restrictions concerning supplies to bidets and other points of draw-off, with byelaw 21(2) permitting certain exceptions.

1. BIDETS WITH OVER-RIM SUPPLY CONNECTED TO A SUPPLY OR DISTRIBUTING PIPE

(a) The requirements of byelaws 19 and 20 would be accepted as being satisfied whenever a bidet is equipped with separate hot and cold taps and site verification of the air gaps confirms they are in accordance with the requirements of the Table in byelaw 16(2).

Such air gaps can be obtained whenever draw-off taps complying with either BS 5412 or BS 5413 are installed on bidets complying with the appropriate requirements of BS 5505: Bidets Part 3, Vitreous china bidets over-rim supply only.

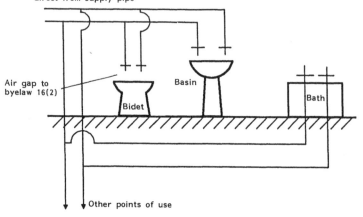

Diagram 57 - Over-rim type bidet

Arrangements complying with diagram 57 permit connection either to hot and cold water distributing pipes or direct to the supply pipe for cold water and to a supply pipe fed unvented heater for hot water. There are no restrictions on the taking of feeds from the distributing or supply pipes to other draw-off taps or water fittings at any level.

    (b) In the case of a bidet equipped with a single outlet (single flow) mixing tap and site verification confirms the air gap is in accordance with the Table in byelaw 16(2) hot and cold supplies can be provided by distributing pipes or direct from the supply pipe as described above. When the water pressures are balanced there are no restrictions on the taking of feeds from either the distributing or supply pipes to other draw-off taps or water fittings at any level.

    If the water pressures are unbalanced, single check valves must be provided in the pipes supplying the mixer tap (see guidance to byelaw 16).

66

(c) In the case of a bidet equipped with a double outlet (divided flow) mixing tap the comments given in (a) for separate hot and cold taps apply.

2. BIDETS WITH ASCENDING SPRAY CONNECTED TO SUPPLY PIPE

The requirements of byelaw 21 would be accepted as being satisfied when a bidet not of the over-rim feed type is connected to an unvented hot water system if hot and cold water are supplied via a break pressure tundish which provides an air gap meeting with the requirements of the Table in byelaw 11 between the supply pipe and the bidet (see diagram 58).

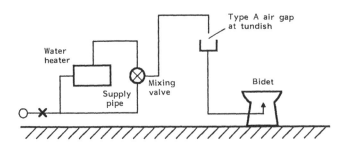

Diagram 58 - Ascending spray bidet

There are no restrictions on the taking of feeds to other draw-off taps or water fittings providing the connections are made upstream of the mixing valve.

NOTE This arrangement could be used when hot and cold water is provided in distributing pipes at balanced pressure.

3. BIDETS WITH ASCENDING SPRAY, CISTERN FED

The requirements of byelaw 21 would also be accepted as being satisfied in the following examples of balanced cistern fed systems.

(a) Water under cistern pressure is conveyed by separate distributing pipes which connect to no other tap or appliance other than the feed pipe to a flushing cistern (see byelaw 21(2)(b)). The

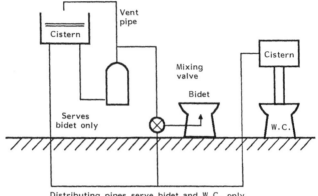

Distributing pipes serve bidet and W.C. only

Diagram 59 - Ascending spray type Bidet

water heater serves the bidet only (see diagram 59).

(b) Water under cistern pressure is conveyed in separate distributing pipes to the bidet. The separate hot water distributing pipe is connected above the cylinder to the vent pipe at point A (see diagram 60) and is fitted with an additional vent pipe connected to it at point B downstream of a check valve (see byelaw 21(2)(c)(i)). Points A and B must be at least 300mm higher than the spill-over level of the bidet.

NOTE   Certain types of check valve have a high headloss and if the cistern is at a low level the available pressure at the bidet may be inadequate to give a satisfactory spray.

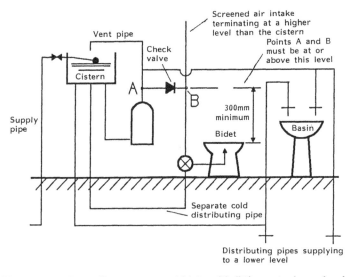

Diagram 60 - Ascending spray type bidet - with fittings at a lower level

In this example there are no restrictions on the taking of supplies to other points of use at any level provided the water is conveyed in distributing pipes not connected to those serving the bidet (other than the cold water to a flushing cistern).

    (c)   Water under cistern pressure is conveyed in distributing pipes to the bidet which is the lowest point of use. The vented hot water distributing pipe and the cold water distributing pipe are both connected at least 150mm higher than the spill-over level of the bidet (see byelaw 21(2)(c)(i), byelaw 16(3)(c) and diagram 61).

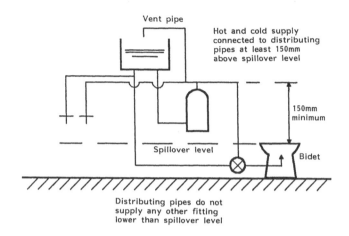

Diagram 61 - Ascending spray type bidet - no fittings at a lower level

There are restrictions on the taking of supplies to other points of use in that none can be any lower than the spill-over level of the bidet.

BYELAWS 22 AND 23

CLOTHES AND DISHWASHING MACHINES

BYELAW 22.

(1) Every clothes washing machine, dishwasher or tumbler drier connected permanently or temporarily to the water service in any premises shall incorporate either a type B air gap or a pipe interrupter which, if removed, renders the machine inoperable.

(2) Subject to paragraph (3), every machine of a kind mentioned in paragraph (1) which is connected permanently or temporarily to the water service elsewhere than in a domestic dwelling shall, in addition to complying with that paragraph, draw water by gravity only from a storage cistern.

(3) Paragraph (2) shall not apply where any machine mentioned in that paragraph incorporates a type A air gap.

70

GUIDANCE BYELAWS 22 AND 23

Built-in protection is required whether a washing machine is permanently or temporarily installed in a dwelling. A type B air gap or a pipe interrupter would be accepted as satisfying this byelaw. Reference should be made to the Directory for a list of acceptable machines satisfying BS 6614 which gives the requirements for the connection of washing machines and dishwashers to water supply pipes. This British Standard is harmonised with a draft European electrical standard (CENELEC HD 274) also covering such connections. The CENELEC document is under review and the British Standard would consequently incorporate any amendments when the standard is published.

A dwelling means any premises, building or part of a building providing accommodation including a terraced house, a semi-detached house, a detached house, a flat or a unit in a block of maisonettes, a bungalow, a flat within any non-domestic premises and any caravan, vessel, boat or houseboat connected to the undertakers' mains.

See also byelaw 88 prescribing permissible water use of washing machines.

The requirements of byelaws 22 and 23 will be accepted as being satisfied in the following cases:

(a) Diagram 62 illustrates a machine installed in a domestic dwelling. If in such a machine a water softener is installed upstream of the integral protective device required by byelaw 22(1) this must be fitted upstream with at least a single check valve which will be accepted as satisfying byelaw 23.

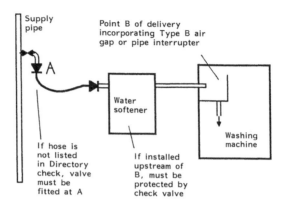

Supply pipe

Point B of delivery incorporating Type B air gap or pipe interrupter

A

Water softener

Washing machine

If hose is not listed in Directory check, valve must be fitted at A

If installed upstream of B, must be protected by check valve

Diagram 62 - Washing Machine in domestic dwelling

(b) Machines installed elsewhere than in a dwelling, satisfying byelaw 22(1) and supplied by gravity from storage. In such cases connections may be made to other fittings and machines from the distributing pipe (see diagram 63).

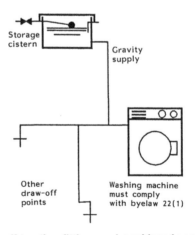

Storage cistern

Gravity supply

Other draw-off points

Washing machine must comply with byelaw 22(1)

Diagram 63 - Draw-off to other fittings and machines in commercial premises

72

(c) Machines installed elsewhere than in a dwelling drawing water by gravity from an interposed cistern supplying only similar machines, subject to (e) below (see diagram 64).

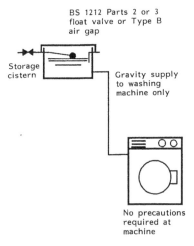

Diagram 64 - No other draw-off from storage so no precautions required at machine in commercial premises

(d) Machines elsewhere than in a dwelling incorporating type A air gaps, subject to (e) below.

(e) Acceptable protection for water softeners upstream of protection required under byelaw 22 for machines in commercial premises would be a double check valve assembly.

BYELAW 24

PIPES CONVEYING WATER TO CISTERNS

BYELAW 24. (1) Every supply pipe conveying water to a cistern (whether or not fitted with a float operated valve) shall -

(a) if the cistern receives or contains, or is likely to receive or contain, any substance which is, or is likely to be, harmful to health, incorporate a type A air gap; or

(b) if the cistern supplies water to a primary circuit in a domestic dwelling, or is a flushing cistern, incorporate either a type B air gap, a pipe interrupter or a double check valve assembly.

(2) Paragraph (1)(b) shall not apply to a supply pipe conveying water to a cistern which -

(a) complies with byelaw 30; or

(b) is fitted with a float operated valve of a reducing flow type which will prevent backsiphonage through it if a vacuum occurs in the feed pipe.

GUIDANCE BYELAW 24

This byelaw requires appropriate protection at the connection of a supply pipe to a cistern to prevent backflow of potentially contaminated water. Other byelaws (33 and 34) deal with the need to control the inflow of water and prescribe normal maximum water levels. See also byelaw 12.

The requirements of byelaw 24 will be accepted as being satisfied in the following circumstances:

(a) Cisterns holding water supplied for domestic purposes and complying with byelaw 30 - no protection required.

(b) Flushing cisterns, feed and expansion cisterns in domestic dwellings or covered cisterns holding water supplied for non-domestic purposes with gravity flow from them at all times - type B air gap, pipe interrupter,

double check valve assembly or a BS 1212 Part 2 or 3 float operated valve installed with its centre line no lower than the centre line of the warning pipe (see diagram 65 and guidance to byelaw 12). The point of interface between the seat and the sealing member will be considered the equivalent datum point to a centre line in any valve with no horizontal seat centre but which satisfies byelaw 24(2)(b).

A float-operated valve of a reducing flow type means a type that progressively reduces the flow of water through it as the level of water in the cistern rises. Modified valves or devices to provide full flow through the valve for a longer period, with a quick shut-off when the water rises to a point of overspill into a separate float chamber, do not afford the backflow protection required under byelaw 24(2)(b) and

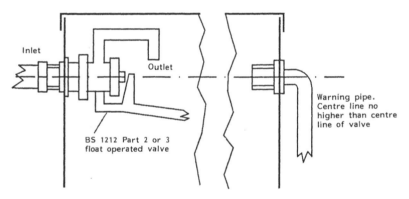

Inlet

Outlet

BS 1212 Part 2 or 3
float operated valve

Warning pipe.
Centre line no
higher than centre
line of valve

Diagram 65 - Installation of BS1212 Part 3 float operated valve in flushing cisterns etc.

additional backflow protection would be necessary.

    (c)   All other cisterns - type A air gap or interposed cistern.

BYELAW 25

GENERAL REQUIREMENTS FOR PROTECTION AT DRAW-OFF POINTS

BYELAW 25.

(1) Every pipe through which water is supplied for domestic purposes to a point of use or draw-off where backflow or backsiphonage is, or is likely to be, harmful to health by reason of a substance which -

(a) is continuously or frequently present in contaminated water, shall incorporate a type A air gap; or

(b) may be present in contaminated water, shall incorporate a type A or type B air gap, a pipe interrupter, a combination of check valve and vacuum breaker, double check valve assembly, or some other no less effective backflow prevention device.

(2) Every pipe through which water is supplied for domestic purposes to a point of use or draw-off where backflow or backsiphonage is not, or is not likely to be, harmful to health, shall incorporate a check valve or some other no less effective backflow prevention device.

(3) Paragraphs (1) and (2) shall not apply where a pipe supplying water to a point of use or draw-off is supplied from a cistern which -

(a) supplies water by gravity only to the point of use or draw-off; and

(b) is installed so that the vertical distance between the spillover level of any vessel containing used or contaminated liquid at any point of use or draw-off, and -

(i) the invert level of the warning pipe in the cistern, is not less than 300mm, and

(ii) the lowest point inside the cistern, is not less than 15mm; and

GENERAL GUIDANCE BYELAW 25

Byelaw 25 identifies 3 levels of backflow risk and the type of protection required at every point of use or draw-off which is supplied with water for domestic purposes. This protection is for points of use that are not specifically covered by byelaws 16 to 24.

The following three schedules list examples of fittings and types of installation which accord with the categories of risk and prescribed protection in byelaws 25(1)(a), 25(1)(b) and 25(2) respectively.

The list is not exhaustive and is given for guidance only. For more information regarding specific manufactured fittings and their backflow protection requirements refer to the Directory.

The protection can be provided in the installation e.g. by an air gap between the outlet of a tap and the rim of a wash basin or in an appliance itself e.g. by an air gap or pipe interrupter built in at the entry to a washing machine.

In an industrial or a commercial establishment or in parts of a research laboratory where it is not necessary to preserve the potability of water and where no water is taken for domestic purposes, protection at every point of use is not required providing supplies to those parts are first taken into a storage cistern through a type A air gap at the inlet.

Byelaw 25(3) details an alternative protection to a type A air gap by interposing a cistern between the supply pipe and the hazard e.g. an industrial process water cistern, but it is essential that the lowest part of the cistern is above the rim of the vessel or other appliance containing contaminated liquid, see diagram 66. For the purposes of this guide the device will be termed "interposed cistern".

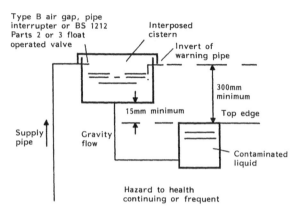

Type B air gap, pipe interrupter or BS 1212 Parts 2 or 3 float operated valve

Interposed cistern

Invert of warning pipe

300mm minimum

Top edge

15mm minimum

Supply pipe

Gravity flow

Contaminated liquid

Hazard to health continuing or frequent

Diagram 66 - Interposed cistern. This is the only acceptable alternative to a type A air gap.

In industrial and other situations the requirements in byelaws 25(1) and (2) for protection at every point of use would not apply beyond such an interposed cistern unless supplies for domestic purposes were drawn from the distributing pipework.

It should be noted that in this latter case there is always a risk of water in the interposed cistern becoming itself contaminated, for example by colouring or disinfecting agents, by dust or cleaning agents. For this reason an appropriate backflow prevention device must also be installed at the inlet to the cistern.

Important notes:

    (a) See byelaw 24(2) where in certain circumstances float operated valves complying with either BS 1212: Part 2 or BS 1212: Part 3 would be acceptable in place of a type B air gap subject to their being installed with their body centre lines no lower in level than the centre line of the warning pipe.

    (b) In the case of a distributing pipe from a cistern supplying several water fittings, protection at point of use is also required to prevent cross-contamination between those fittings.

This is illustrated in diagram 67.

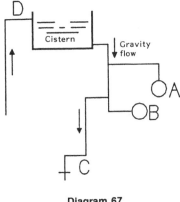

Diagram 67

The cistern serves 3 points of use A, B and C and prevents contamination reaching the cistern feed pipe D. There could be cross-contamination between A to B or C and from B to C unless A and B also incorporate appropriate protection. Point of use protection is not required at C if it is the lowest point of draw-off.

SCHEDULES OF COMPLIANCE

SCHEDULE A

Examples of points of use or delivery of water where backflow is or is likely to be harmful to health from a substance continuously or frequently present (byelaw 25(1)(a)). This complements requirements set out in byelaws 16 to 24.

NOTE   The required protection is a type A air gap at the point of use or an interposed cistern.

General: At the inlet to any industrial cistern from which the flow is not always by gravity.
WC pans.
Urinals.
Bidets (see detailed arrangements set out earlier in Part III).
Non-domestic hose union taps.
Non-domestic central heating systems.
Non-domestic sealed heating systems (see guidance to byelaw 14).
Mobile plant and tankers e.g. gully emptiers, etc.

Fire sprinkler systems containing anti-freeze (see also diagrams 35 to 42, byelaw 12).
Commercial clothes washers and tumbler driers.
Water softening plant except salt regenerated.

Medical:
Bedpan washers.
Laboratories (see also note under comment to byelaw 12, non-domestic supplies within hospitals).
Dental spittoons and equipment.
Mortuary equipment.

Food processing:
Food and vegetable processing plants.
Dairies.
Slaughterhouse equipment.
Butchery and meat trades.

Catering:
Bottle washing apparatus.
Drink vending or dispensing machines in which pressure is raised by a pump delivering more than 10 litres/minute.
Commercial dishwashers.
Refrigerating equipment - water cooled.

Industry and Commerce:
Industrial chemical mixing devices, baths, vats, etc. and generally water taken for processes.
Water treatment plant.
Dyeing equipment.
Sewage treatment and sewerage cleansing plant.
Drain cleaning plant.
Car washing and degreasing plants.
Industrial disinfection equipment.
Printing and photographic equipment.
Commercial clothes washing plant.
Brewery and distillation plant.

Storage:
Storage cisterns connected to non-domestic central heating systems.
Firefighting reserve storage (see advice given on byelaw 12).
Agricultural storage (uncovered or supplemented from non-potable sources).

SCHEDULE B

Examples of points of use or delivery of water where backflow or backsiphonage is or is likely to be harmful to health from a substance which may be present (byelaw 25(1)(b)).

NOTE   Except where otherwise stated, the required protection is a type A or type B air gap, a pipe interrupter, a combination of a check valve and a vacuum breaker or a double check valve assembly.   If an appropriate upstand cannot be provided a combination of check valve and vacuum breaker is unacceptable, i.e. vacuum breakers do not afford protection against backpressure backflow.   Also, if the vacuum breaker is of the "atmospheric" type no control valve is permissible downstream.

Commercial softening plant (common salt regenerated).
Domestic clothes and dishwashing machines and tumbler driers.
*Drink vending and dispensing machines (note that the requirements in cases where CO_2 is injected in such appliances are under review. See inclusion of machines with pumps in Schedule A).
Home dialysing machines (with integral membrane washing).
Domestic feed and expansion cisterns.
Fire sprinkler systems (see under byelaw 12).
Standpipes at marinas serving small boats only.
Standpipes on water mains (other than used for firefighting purposes).
Supplying water to mobile clean water apparatus.
Bar top glass washers.
Cisterns installed under byelaw 24(1)(b).
*NOTE At the time of publication, if CO_2 is injected and there is copper pipe upstream of the point of injection, a double check valve with intermediate vent would be accepted.

SCHEDULE C

Examples of points of use or delivery of water where backflow or backsiphonage is not or is not likely to be, harmful to health (byelaw 25(2)).

NOTE   A check valve would be the usual protection, but an atmospheric type vacuum breaker would be accepted if properly installed (see byelaw 11 Section 6) and there is no possiblity of backpressure backflow.

Single outlet hot and cold water mixing taps in domestic premises (unbalanced systems) - to prevent hot water entering supply pipe: protection additional to protection at point of discharge of mixed water.

Home dialysing machines (without integral membrane washing).
Domestic softening plant (common salt regenerated).
Fire sprinkler systems in some cases (see under byelaw 12).
Drink vending machines in which no ingredients or CO_2 are injected into the supply pipe.

Other backflow prevention devices under consideration

The devices specifically mentioned in byelaws 25(1) and (2) are described under byelaw 11 but byelaw 25 also mentions "other no less effective devices". As stated elsewhere in this guide such devices will be tested by the UK Water Fittings Byelaws Scheme and if found acceptable will be listed in the Directory under the appropriate category of protection. The undertakers should be consulted if any unlisted device is proposed to be used.

Devices under investigation include:

Terminal anti-vacuum valves (BS 6282: Part 2).
In-line anti-vacuum valves (BS 6282: Part 3).
Combined check and anti-vacuum valves (BS 6282: Part 4).
Reduced pressure principle backflow preventers (ASSE 1013).
Pipe applied atmospheric type vacuum breakers (ASSE 1001).
Double check valve type back pressure backflow preventers (ASSE 1015).
Hose connection vacuum breakers (ASSE 1011).
Vacuum breakers anti-siphon pressure type (ASSE 1020).
Backflow preventers with intermediate atmospheric vent (ASSE 1012).
Pipe disconnector to German standard DIN 3266 Part I.

The ASSE references are to documents produced by the American Society of Sanitary Engineering, P O Box 40362, Bay Village, Ohio 44140, USA. The British standards mentioned are under review.

BYELAW 26

SECONDARY BACKFLOW PROTECTION

BYELAW 26. Every -

(a) supply or distributing pipe which conveys water to two or more separately occupied premises (whether or not they are separately chargeable by the undertakers for a supply of water); and

(b) supply pipe which conveys water to premises which under any enactment are required to provide a storage cistern capable of holding sufficient water for not less than 24 hours ordinary use;

shall be fitted with such a combination or combinations of check valves, vacuum breakers, double check valve assemblies or some other no less effective backflow prevention device, as will effectively prevent the backflow or backsiphonage of water from one of those premises to another.

GUIDANCE BYELAW 26

Other byelaws deal with protection of the supply and with protection at point of use. In certain installations however the risks are increased because of the possibility of internal backflow within installations in buildings in multiple occupation.

In these cases this byelaw requires that certain additional precautions should be taken in the event of low pressure to prevent or limit backflow in certain supply and distributing pipes. These include the main supply pipes or distributing pipes serving several dwelling units such as in a block of flats where contaminated water might pass from one flat to another. The byelaw exempts any separately occupied premises with its own supply pipe.

The precautions would also limit the risks in industrial premises, schools, offices, hospitals, etc. and although the byelaw does not apply to such premises, secondary backflow protection is strongly recommended to be installed, floor by floor, in them.

The requirements of byelaw 26 should be applied in the following cases:

- on the common supply pipe system serving two or more houses, flats or maisonettes and on the common distributing pipe system serving two or more such premises.

- on the supply pipes of individual dwellings arranged to receive an intermittent supply of water.

Section (a) below deals with protection associated with supply pipes and Section (b) deals with that associated with distributing pipes. In each case the requirement of the byelaw would be accepted as being satisfied if the measures indicated are taken.

It is not practicable to provide secondary backflow protection to recirculatory hot water systems except in cases where the hot water feed pipe to an individual premises is not itself a recirculatory flow and return system.

Except in this latter event, secondary backflow protection would not therefore be required but especial care should be taken to ensure full conformity with byelaw 25. Wherever practicable appliances should be down-fed from ceiling level and high risk appliances (e.g. bedpan washers) should be fed from dedicated storage.

(a) Protection associated with supply pipes

It is recommended that secondary backflow protection be installed at every floor level as indicated in diagram 68. Protective devices which would be accepted would consist of double check valve assemblies installed immediately downstream of the stop valve on each branch supply pipe to the level or floor under consideration.

Note that because no control valve can be installed downstream of an atmospheric type vacuum breaker, such a device cannot be used in these circumstances. The possible use of a combined check valve and pressure type vacuum breaker is being investigated.

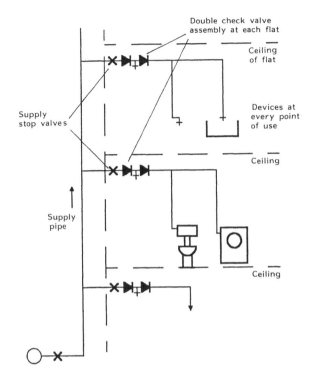

Diagram 68 - Secondary backflow protection of supply pipes

(b) Protection associated with distributing pipes

The following describes an acceptable arrangement based on a vented system (see diagram 69):

(1) A vent pipe is connected and arranged on each distributing pipe at its junction with the associated storage cistern. Every vent pipe should be the same size as the distributing pipe at the point of connection to the cistern; and

(2) Every water fitting or appliance is only connected to a distributing pipe by means

of a branch pipe. No part of any branch pipe is at a higher elevation than the point of its junction with the distributing pipe; and

(3) Every branch pipe is arranged in such a way that the overflowing level of associated fixed appliances served by a draw-off fitting connected to the branch pipe is at a level not less than 300mm below the point of the junction of the branch pipe with the falling length of distributing pipe.

Provided that requirements (b)(2) and (b)(3) need not apply to any fixed appliance which is installed so that its overflowing level is at the lowest elevation within any distributing pipe system.

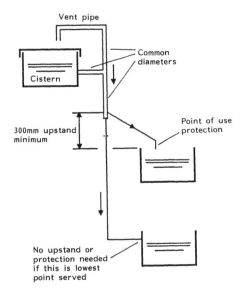

Diagram 69

Examples of secondary backflow protection on distributing pipes

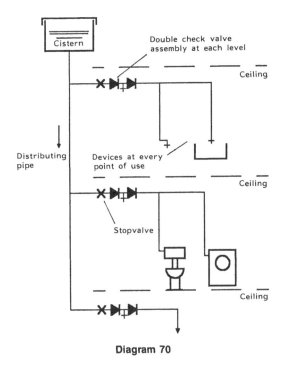

Diagram 70

Alternative secondary backflow protection on distributing pipes.

As an alternative to the arrangement described above, additional devices such as a double check valve assembly could be fitted at each floor level as shown in diagram 70.

BYELAW 27

PIPES TO BE READILY DISTINGUISHABLE

BYELAW 27. In any premises, other than a domestic dwelling, every -

(a) supply pipe; and

(b) pipe for supplying water solely for firefighting purposes;

shall be clearly and indelibly marked so that such pipes are readily distinguishable from each other and every other pipe in those premises.

The purpose of this byelaw is to warn against accidental cross-connections that could lead to contamination of supply pipes. In industrial and commercial establishments, etc. byelaw 25 does not apply where the proviso in that byelaw is complied with. It is therefore necessary clearly to distinguish these parts of the installation from those to which byelaw 25 applies. This is a matter not only of "byelaw" concern but also a measure contributing to the health and safety of persons within these establishments. Plastics pipes underground will be accepted as satisfying this byelaw if they are coloured or pigmented blue.

"Readily distinguishable" means that any method of identification or marking would be accepted whereby only visual examination will be necessary to enable it to be identified as a supply pipe. Methods include colour pigmentation incorporated in plastics pipes or colour painting of pipes and fittings or permanent marks or labels or, above ground, as specified in BS 1710: Identification of pipelines and services.

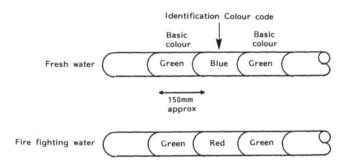

Diagram 71 - Examples of BS 1710 colour identification codes

NOTE The colour identification should be placed at junctions and at both sides of valves, service appliances and bulkheads.

BYELAW 28

SEPARATION OF PIPES IN FIREFIGHTING INSTALLATIONS
FROM OTHER FITTINGS

> **BYELAW 28.** No water fitting shall be connected to any pipe installed solely for the supply of water for firefighting purposes except a water fitting or other equipment installed solely for those purposes.

GUIDANCE BYELAW 28

Because water not supplied by the undertaker and of uncertain quality can be pumped into fire mains installations, it is necessary to prevent interconnection between such installations and supply and distributing pipes.

See diagrams 35 to 42 in guidance to byelaw 12 for acceptable arrangements for taking supplies for sprinkler installations.

BYELAW 29

ACCESSIBILITY OF BACKFLOW PREVENTION DEVICES

> **BYELAW 29.** Every backflow prevention device shall, so far as is reasonably practicable, be installed so that it is accessible for examination, repair or replacement.

GUIDANCE BYELAW 29

It is essential that to ensure satisfactory operation, backflow prevention devices with moving parts should be maintained regularly. They should therefore be readily accessible for maintenance without the use of special tools or removing floor boards or wall finishes, etc.

PART IV

PREVENTION OF WASTE OR CONTAMINATION OF STORED WATER

CONTENTS

PART IV - PREVENTION OF WASTE OR CONTAMINATION OF STORED WATER

BYELAW 30

CISTERNS STORING WATER FOR DOMESTIC PURPOSES

BYELAW 30.   (1) Every storage cistern for water supplied for domestic purposes, shall -

    (a) be installed in a place or position which will prevent the entry into that cistern of surface or groundwater, foul water, or water which is otherwise unfit for human consumption; and

    (b) comply with paragraph (2).

(2) Every cistern of a kind mentioned in paragraph (1) shall -

    (a) be insulated against heat and frost; and

    (b) when it is made of a material which will, or is likely to, contaminate stored water, be lined or coated with an impermeable material designed to prevent such contamination;

    (c) have a rigid, close fitting and securely fixed cover which -

      (i) is not airtight,

      (ii) excludes light and insects from the cistern,

      (iii) is made of a material or materials which do not shatter or fragment when broken and which will not contaminate any water which condenses on its underside,

      (iv) in the case of a cistern storing more than 1 000 litres of water, is constructed so that the cistern may be inspected and cleansed without having to be wholly uncovered, and

GUIDANCE BYELAW 30(1)

The requirements of byelaw 30(1) would be accepted as being satisfied
whenever a storage cistern, covered in accordance with the
requirements of byelaw 30(2):

(a) is placed wholly above ground level, preferably
on a well drained site, not liable to flooding,
otherwise such that its base is not less than
600mm above the highest known flood level; or

(b) being a concrete reservoir, and is buried or
partly sunk in the ground, has been designed,
constructed and tested in accordance with BS
5337: Code of Practice for the structural use of
concrete for retaining aqueous liquids; or

(c) is located in a watertight basement below
ground level.

GUIDANCE BYELAW 30(2)

A new British Standard is in course of preparation for a cistern which
would comply with byelaw 30(2). Diagram 72 illustrates the principles
of the byelaw to be incorporated in the Standard. Such a cistern would
be accepted as satisfying the byelaw. Where a cistern storing water
for domestic purposes is being replaced, the replacement shall comply
with this byelaw.

To restrict microbiological growth it is important that stored potable
water should be kept at as low a temperature as practicable, ideally
less than 20°C. Insulation provided against frost should be beneficial
in this respect.

Immersion heaters will be accepted for use in cold water storage
cisterns if the heat is solely provided for frost protection purposes. The

heat source must be automatically controlled and must not be capable of raising the temperature to more than sufficient for the purpose.

In the case of a feed cistern element of a hot water storage combination unit the temperature is permitted in the British Standard (BS 3198) to rise to not more than 39°C. Such a cistern would be accepted as meeting the requirements of byelaw 30(2)(a) if:

> (a) the unit only supplies hot water for domestic purposes,
>
> (b) the cistern is incorporated as an integral part of the unit,
>
> (c) no water is drawn from it except to the hot water storage vessel,
>
> (d) it is effectively insulated against frost, and
>
> (e) in all other respects it meets with the requirements of the byelaws.

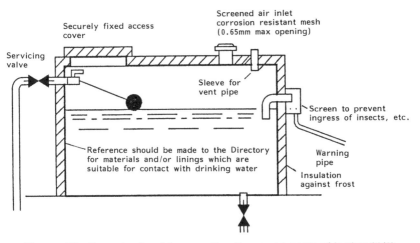

Diagram 72 - Example of a cistern meeting the requirements of byelaw 30(2)

The purpose of byelaw 30(2)(iv) is to ensure that covers of cisterns are not so large as to discourage maintenance and adjustment of float operated valves or inspection and cleansing of the cistern.

It will be accepted as being satisfied in larger cisterns if an area of cover can be readily removed to permit entry for inspection and cleansing and a smaller area can be removed for maintaining the float operated valve.

BYELAW 31

PLACING OF STORAGE CISTERNS

BYELAW 31. Every storage cistern shall be installed in a place or position such that -

(a) the inside may be readily inspected and cleansed; and

(b) any float-operated valve or other device used for controlling the inflow of water may be readily installed, repaired, renewed or adjusted.

GUIDANCE BYELAW 31

The requirements of byelaw 31 would be accepted as being satisfied where a storage cistern has an unobstructed space above it of not less than 350mm. In the case of deep and narrow cisterns, for example hot water combination units, access for cleaning via a handhole may be required. The handhole should be circular or elliptical of area not less than 7 850mm^2 and its centre should be not more than 125mm above the lowest horizontal surface of the cistern or unit. In the case of small cisterns, the overhead unobstructed space may be reduced to 225mm provided no dimension of the cistern exceeds 450mm in any plane (see diagram 73).

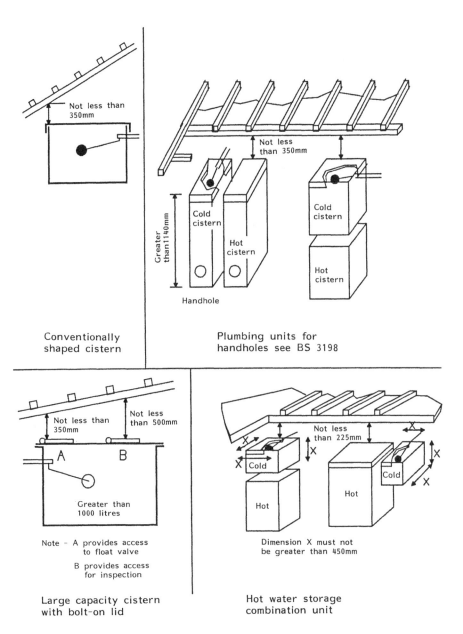

Not less than
350mm

Conventionally
shaped cistern

Not less
than 350mm

Greater than1140mm

Cold
cistern

Hot
cistern

Handhole

Cold
cistern

Hot
cistern

Plumbing units for
handholes see BS 3198

Not less than
350mm

Not less
than 500mm

A   B

Greater than
1000 litres

Note - A provides access
to float valve

B provides access
for inspection

Large capacity cistern
with bolt-on lid

Not less
than 225mm

X

X   Cold

Hot

X

Cold

X

Hot

Dimension X must not
be greater than 450mm

Hot water storage
combination unit

Diagram 73 - Clear space needed above storage cisterns

BYELAW 32

SUPPORT OF STORAGE CISTERNS

GUIDANCE BYELAW 32

The requirements of byelaw 32 would be deemed to be satisfied if a storage cistern made of plastics, fibreglass, etc. is supported on close boarded timber laid on support beams (see diagram 74).

Where a cistern is made of rigid materials, e.g. galvanised steel, it can be placed directly on support beams spaced at not more than 350mm centres.

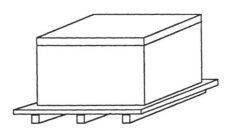

Diagram 74 - Example of close boarded support to plastics cistern

98

BYELAW 33

PIPES SUPPLYING WATER TO STORAGE CISTERNS

BYELAW 33.

(1) Every pipe supplying water to a storage cistern shall be fitted with a float-operated valve or some other no less effective device for controlling the inflow of water by preventing any overflow.

(2) Paragraph (1) shall not apply to a pipe connecting two or more storage cisterns each of which has the same overflowing level.

GUIDANCE BYELAW 33

Byelaw 33 requires inflow to cisterns to be properly controlled. The most straightforward system is control by a float-operated valve but there are other suitable methods based on shut-off by level sensors controlling remote electrically or pneumatically operated valves. The undertakers will be able to advise as to their acceptability and reference should be made to the Directory.

The proviso set out in byelaw 33(2) is illustrated in diagram 75.

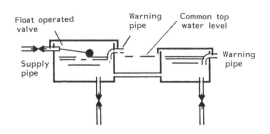

Diagram 75 - Example of byelaw 33(2) - Connecting cisterns

NOTE This arrangement is not recommended because of the possibility of stagnation of water in the second cistern. It is preferable to have a float operated valve supplying each cistern.

BYELAW 34

FIXING AND ADJUSTMENT OF FLOAT-OPERATED VALVES

BYELAW 34.   (1) Every float-operated valve or other device which controls the inflow of water to a storage cistern shall be -

(a) securely and rigidly fixed to that cistern; and

(b) installed so that the inflow of water is shut off when the level of the water in the cistern -

(i) is not less than 25mm below the overflowing level of that cistern, or

(ii) where the cistern is fitted with a device mentioned in byelaw 38(2) below, is not less than 50mm below the overflowing level of that cistern.

(2) Every feed pipe supplying water to such a valve or other device as is mentioned in paragraph (1) shall be connected, braced and supported so as to prevent it from moving or buckling in relation to the thrust of that valve or other device.

GUIDANCE BYELAW 34

The purpose of byelaw 34(1)(a) and 34(2) is to ensure that at all times:

(a) the setting of the float-operated valve is maintained and the highest water level is not increased by the yielding of the material of the cistern or of any bracket, and

(b) load is not transferred to the supply pipe.

These requirements are illustrated in diagram 76.

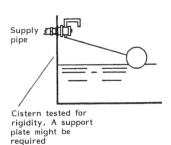

Supply pipe

Cistern tested for
rigidity. A support
plate might be
required

Float operated valve
fixed to cistern wall

Bracket or plate
of adequate strength
to withstand thrust
of float arm

Float operated valve
fixed to a bracket.
E.g. in a Type A
air gap

Diagram 76 - Securing of float-operated valves

Byelaw 34(1)(b) requires that whenever a new cistern is installed or any
work is carried out on the float-operated valve the shut-off level must
be adjusted to provide an air gap below the invert of the warning pipe
or lowest overflow pipe.   This byelaw is related to byelaw 38 which sets
out the requirements for warning and overflow pipes.

Note also byelaw 69(1) concerning servicing valves, and the guidance
to byelaw 24 dealing with backflow protection and also byelaw 92
dealing with allowance for expansion water. Diagrams 77 to 80 illustrate
the setting of float-operated valves and warning devices in relation to
cistern size.

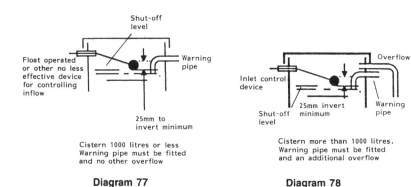

Shut-off
level

Float operated
or other no less
effective device
for controlling
inflow

Warning
pipe

25mm to
invert minimum

Cistern 1000 litres or less
Warning pipe must be fitted
and no other overflow

Diagram 77

Overflow

Inlet control
device

25mm invert
minimum

Shut-off
level

Warning
pipe

Cistern more than 1000 litres.
Warning pipe must be fitted
and an additional overflow

Diagram 78

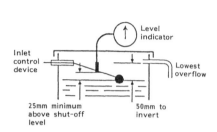

Cistern more than 5000 litres but not
more than 10 000 litres. Overflow must
be fitted but warning pipe may be omitted.

Diagram 79

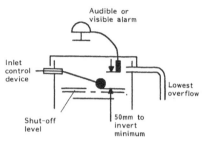

Cistern more than 10 000 litres. Overflow
must be fitted but warning pipe may be
omitted.

Diagram 80

BYELAWS 35, 36 and 37

PREVENTION OF CONTAMINATION FROM PRIMARY HEATING
CIRCUITS AND FEED CISTERNS

BYELAW 35.   (1) No vent pipe connected to a cistern-fed primary circuit shall be installed to convey water into a cistern which supplies water to a secondary system.

           (2) No warning or overflow pipe from any cistern connected to a vented primary circuit shall be installed to convey water to any cistern from which water may be drawn for domestic purposes.

BYELAW 36.   (1) In every double feed indirect cylinder from which stored hot water is, or may be, drawn for domestic purposes, the pressure in the primary heater within that cylinder shall not exceed the pressure of the stored water under normal operating conditions.

           (2) Paragraph (1) shall not apply to a cylinder inside which the primary heater -

           (a) has no joints; or

           (b) is constructed so that any joints will withstand any water pressure to which they are, or may be, subject under normal operating conditions.

102

GUIDANCE BYELAW 35

Byelaw 35(1) prevents contamination of a domestic storage cistern from a primary circuit vent pipe. Diagram 81 shows the correct and incorrect positions of the latter.

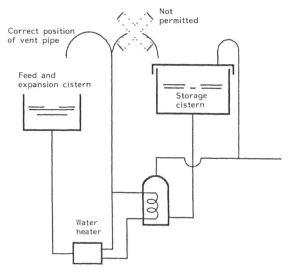

Diagram 81

103

Byelaw 35(2) prevents contamination of a domestic storage cistern from a primary feed and expansion system. Diagram 82 illustrates incorrectly installed cisterns. Note that the feed cistern should not be located at a higher level than the storage cistern.

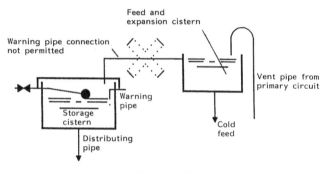

Diagram 82

GUIDANCE BYELAW 36

Byelaw 36 prevents contamination of the secondary hot water system arising out of a leak in the primary coil in an indirect cylinder. The pressure in the primary circuit must normally be not more than that in the secondary and the construction of the coil must be satisfactory. Byelaw 36(2) will be accepted as being satisfied in respect of any hot water cylinder with a coil type primary heater which complies with BS 1566: Copper indirect cylinders for domestic purposes: Part 1 Double feed indirect cylinders.

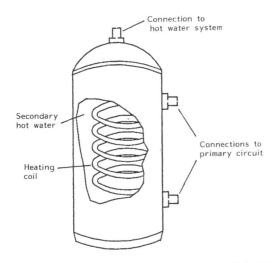

Connection to
hot water system

Secondary
hot water

Heating
coil

Connections to
primary circuit

Diagram 83 - Example - Coil heated indirect cylinders (BS 1566: Part 1)

GUIDANCE BYELAW 37

Byelaw 37 prevents contamination of the secondary hot water within a single feed indirect cylinder in which an air pocket separates it from the primary heating water. These cylinders must be supplied from storage and vented. No corrosion inhibitors or additives are permitted. The operation of such a cylinder is illustrated in diagram 84.

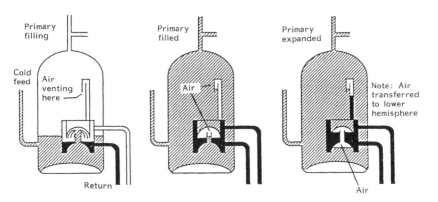

Primary
filling

Cold
feed

Air
venting
here

Return

Primary
filled

Air

Primary
expanded

Note: Air
transferred
to lower
hemisphere

Air

Diagram 84 - Filling sequence of a single feed indirect cylinder

105

The requirements of byelaw 37 would be accepted as being satisfied where a single feed indirect cylinder complies with BS 1566: Copper indirect cylinders for domestic purposes: Part 2 Single feed indirect cylinders, subject to any associated primary circuit having a water volumetric capacity not more than the maximum specified in that BS for the particular cylinder. A correct installation of such a cylinder is illustrated in diagram 85.

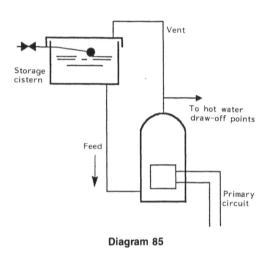

Diagram 85

BYELAWS 38 to 41

WARNING AND OVERFLOW PIPES

BYELAW 38.   (1) Every storage cistern which has a capacity exceeding 1 000 litres shall, subject to paragraph (2), be fitted with an overflow pipe and a warning pipe, and every other storage cistern shall be fitted with a warning pipe only.

           (2) Paragraph (1) shall not apply to require the fitting of both an overflow pipe and a warning pipe where -

(a) in the case of a storage cistern with a capacity exceeding 5 000 litres but not exceeding 10 000 litres, that cistern is fitted with an instrument which indicates when the water level is not less than 25mm below the overflowing level of the lowest overflow pipe; or

(b) in the case of a storage cistern with a capacity exceeding 10 000 litres, that cistern is fitted with an audible or visual alarm operating independently of the valve or device which controls the inflow of water and indicates when the water in the cistern is about to overflow.

(3) In this byelaw "capacity" means the volume of water which the cistern is capable of holding measured to its overflowing level.

BYELAW 39. Every warning pipe shall be installed so as to discharge water immediately the water in the cistern reaches overflowing level.

BYELAW 40. No warning or overflow pipe shall comprise, include or have connected to it, any flexible hose.

BYELAW 41. Where two or more cisterns have a common warning pipe that pipe shall be installed so that the source of any overflow may be readily identified and shall be so arranged that any overflow from the cistern cannot discharge into another.

GUIDANCE BYELAW 38

Warning pipes are the sole permitted overflow pipes in cisterns up to 1 000 litres capacity. Above this size overflow pipes must be fitted in addition and the warning pipe may be omitted in cisterns above 5 000 litres capacity providing other arrangements are made for giving warning.

The byelaws have no requirement concerning the size of a warning pipe but it is implied that it shall be large enough so that the discharge is conspicuous. BS 6700 recommends that the size shall be no less than 22mm nominal diameter (or 19mm bore).

In the case of cisterns of capacity between 5 000 and 10 000 litres a level indicator which operates when the water reaches not less than 25mm below the invert of the overflow is sufficient. In the case of a larger cistern a visual or audible alarm operating independently of the inlet control device is required.   This latter provision would not be required if, in addition to the float valve or other device controlling the inflow, there is an independent control to shut off the inflow when the water level reaches the overflowing level of the warning pipe or lowest overflow pipe as the case may be.

These arrangements are illustrated in guidance to byelaw 34 and in diagram 86 which illustrates a case where flow control is by remote control of a pump.

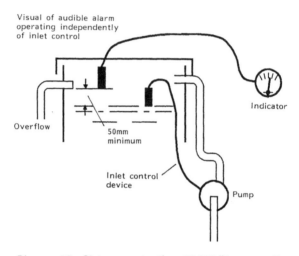

Diagram 86 - Cistern greater than 10 000 litres capacity

GUIDANCE BYELAW 39

Byelaw 39 is to ensure that the warning pipe actually discharges water immediately water reaches the overflowing level.

GUIDANCE BYELAW 40

Byelaw 40 requires that warning pipes must be made of rigid material and no hose is attached that might prevent overflow from being conspicuous.   It would be accepted as being satisfied if any material listed in Part VI is used.

Warning pipes should discharge in a conspicuous position. The following are examples of acceptable arrangements:

(a) In dwellings. The warning pipe discharges outside an external wall or it discharges not less than 150mm above the rim of a lavatory pan.

(b) Common warning pipe. Each warning pipe discharges into a tundish in a visible position such that there is a clear air gap between the point of discharge and the rim of the tundish. The tundish itself drains into a common warning pipe discharging outside an external wall.

(c) Exceptionally, e.g. in hospitals, etc., the undertakers may accept a warning pipe discharging via a clear air gap into a tundish which may have no visible outlet.

GUIDANCE BYELAW 41

Byelaw 41 permits the joining together of two or more lengths of pipe to form a common warning pipe providing the discharge is readily seen and the combined overflow is designed so that the overflow from one cistern cannot discharge into another. From both the consumer's and undertakers' point of view, outlets which discharge outside a building are best. However, where this is not practicable owing to the design of the building, overflow warning pipes may be discharged internally provided the discharge is conspicuous. The use of a property designed combination bath overflow manifold would be accepted as meeting the requirements (see diagram 87).

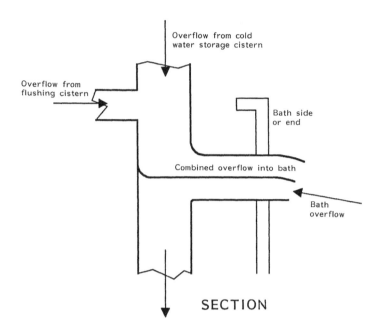

Overflow from cold
water storage cistern

Overflow from
flushing cistern

Bath side
or end

Combined overflow into bath

Bath
overflow

SECTION

Diagram 87 - Principle of operation of combined bath overflow manifold

NOTE   In Scotland byelaw 41 reads:

"Where two or more cisterns have a common warning pipe, that pipe shall be installed so that the source of any water discharging from it can be readily identified."

BYELAW 42

REQUIREMENTS FOR FLOAT-OPERATED VALVES

BYELAW 42.   Every float-operated valve installed in any cistern or other apparatus shall -

(a) be capable of controlling the flow of water into that cistern or apparatus; and

(b) when it is closed, be watertight; and

(c) incorporate either a renewable seat and washer which are resistant to both corrosion and erosion by water or some other no less effective valve assembly; and

(d) on installation, be capable of withstanding without leaking when closed an internal hydraulic pressure 1.5 times the pressure to which it will ordinarily be subject; and

(e) have a float which -

   i) is constructed of a material capable of withstanding without leaking any water temperature in which it operates or is likely to operate, and

   ii) has a lifting effort such that when not more than half immersed, the valve is capable of droptight closure against the highest pressure to which that valve is likely to be subject; and

(f) have a lever which -

   i) when the valve is closed will withstand without bending or distorting a force twice that to which it is ordinarily subject, and

   ii) in the case of a $^1/_2$ inch valve, is constructed so that the water shut-off level may be altered or adjusted without bending the float lever.

The requirements of byelaw 42 in respect of any float-operated valve and its operating lever (excluding its float) will be accepted as being satisfied if it complies with:

BS 1212:   Float-operated valves (excluding floats): Part 1 Piston type.

NOTE   Excepting $^1/_2$ inch size.

BS 1212:   Float-operated valves (excluding floats): Part 2 Diaphragm type (brass body).

BS 1212:   Float-operated valves (excluding floats): Part 3 Diaphragm type (plastics body) for cold water services.

The Part 2 and 3 valves are capable of withstanding a backsiphonage test when the water level is as high as the centre line of the valve whereas a Part 1 valve will not withstand such a test.

NOTE   There are many float-operated valves available which do not comply with BS 1212 and therefore reference should be made to the Directory for other acceptable types.   Care should be taken to ensure that they are installed so as to satisfy the requirements of byelaw 24.

The requirements of byelaw 42 would be accepted as being satisfied if any float is of the appropriate size having regard to the working pressure and it complies with either:

        (a) BS 1968: Floats for ball valves (copper), or

        (b) BS 2456: Floats (plastics) for ball valves for hot and cold water.

Byelaw 42 requires not only that the lever of a float-operated valve should be strong enough but that in the case of every $^1/_2$ inch sized valve the float should be fitted with some easily adjustable device for setting the waterlevel.   Bending of the float arm will not be accepted as a means of meeting the latter provision and BS 1212: Part 2 illustrates an acceptable method (see diagram 88).

NOTE   A $^1/_2$ inch size float-operated valve to BS 1212: Part 1 does not meet this requirement.

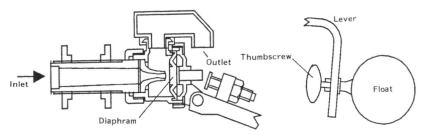

Diagram 88 - Details of float-operated valves

BYELAW 43

FLOAT-OPERATED VALVES CONVEYING HOT WATER

> **BYELAW 43.** No float-operated valve shall be installed to convey hot water to any cistern unless -
>
> (a) it is constructed of materials capable of withstanding without leaking any ordinary operating water temperature to which it is or may be subject; and
>
> (b) so far as it is reasonably practicable, its operation is not, and is not likely to be, prevented or impaired by scale; and
>
> (c) having regard to any scale which is, or is likely to be, deposited on the valve or float, it is adjusted to prevent any overflow.

GUIDANCE BYELAW 43

BS 2456 floats are not designed for continuous contact with hot water and copper floats for such a purpose should have brazed or equivalent seams. Float-operated valves conveying hot water must not incorporate materials unsuitable for continuous or frequent contact with hot water; BS 1212 does not include relevant requirements.

This byelaw is not intended to apply to float-operated valves conveying cold water to feed cisterns supplying hot water apparatus, except that the float must be capable of withstanding an occasional rise in temperature due to any expansion water being received in the cistern.

BS 2456 and BS 1968 floats will be accepted as satisfying this requirement.

BYELAW 44

FLOW CONTROL DEVICES OTHER THAN FLOAT-OPERATED VALVES

> BYELAW 44. Every valve or device installed for controlling the inflow of water into any storage cistern (other than a float-operated valve) shall be capable of controlling the flow of water into that cistern.

GUIDANCE BYELAW 44

See guidance to byelaw 33.

BYELAW 45

LOCATION OF STORAGE CISTERNS SUPPLYING WATER FOR NON-DOMESTIC PURPOSES

> BYELAW 45. No storage cistern for water supplied for non-domestic purposes and no cylinder or water fitting used in connection with it shall be installed in a place or position which may result in the stored water becoming unfit for the purpose for which it is intended.

GUIDANCE BYELAW 45

Byelaw 45 applies to the placing of cisterns storing water in environments where there are risks of contamination of the stored water by ingress of external sources of contaminants. The requirement will be accepted as being satisfied whenever any cistern is placed in a loft, room or other enclosed space in which no industrial or like activity takes place, or materials stored, which might result in air pollution or accidental spillages or contaminating materials gaining access to the stored water. The quality of water must also be preserved by suitably designed rigid covers, etc.

BYELAW 46

ANIMAL DRINKING TROUGHS AND BOWLS

GUIDANCE BYELAW 46

The requirements of this byelaw would be accepted as being satisfied if any animal watering trough complies with BS 3445: Fixed agricultural water troughs and water fittings.

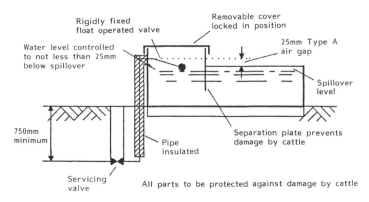

Rigidly fixed float operated valve

Removable cover locked in position

Water level controlled to not less than 25mm below spillover

25mm Type A air gap

Spillover level

750mm minimum

Separation plate prevents damage by cattle

Pipe insulated

Servicing valve

All parts to be protected against damage by cattle

Diagram 89 - Example of a cattle trough

115

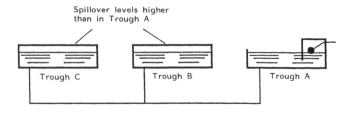

Spillover levels higher
than in Trough A

Trough C   Trough B   Trough A

Diagram 90 - Example of supplying two troughs from another

Troughs B and C are arranged at such a level that any overflow takes place at A.   Trough A arranged as in diagram 89.

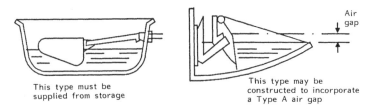

This type must be
supplied from storage

Air
gap

This type may be
constructed to incorporate
a Type A air gap

Diagram 91 - Examples of animal drinking bowls

BYELAW 47

PONDS, FOUNTAINS AND POOLS

BYELAW 47.   Every pond, fountain or pool the capacity of which exceeds 10 000 litres and which is filled or supplied with water from the undertakers' mains shall have an impervious lining or membrane to prevent the leakage or seepage of water.

GUIDANCE BYELAW 47

This byelaw would be accepted as being satisfied if any reservoir is constructed in accordance with BS 5337: Code of Practice for the structural use of concrete for retaining aqueous liquids.

PART V
PREVENTION OF WASTE
OF WATER FROM
DAMAGE TO WATER
FITTINGS FROM CAUSES
OTHER THAN
CORROSION

CONTENTS

PART V - PREVENTION OF WASTE OF WATER FROM DAMAGE TO WATER FITTINGS FROM CAUSES OTHER THAN CORROSION

BYELAW 48

COVERING OF PIPES

BYELAW 48.   (1) Subject to paragraphs (2) and (3) below, the vertical distance between the top of every pipe or other water fitting laid or installed below ground and the finished ground level shall be -

(a) not less than 750mm; and

(b) not more than 1.35m.

(2) Where it is impracticable to comply with sub-paragraph (1)(a), a pipe or other water fitting shall be laid or installed as deep as is reasonably practicable below the finished ground level and shall be effectively protected against damage from freezing and from any other cause.

(3) This byelaw shall not apply to any pipe or other water fitting which is laid or installed in the ground under any building or structure of a permanent nature.

GUIDANCE BYELAW 48

Diagram 92 shows acceptable examples of laying pipes in normal conditions and over or under obstructions. For details of insulation see guidance to byelaw 49.

119

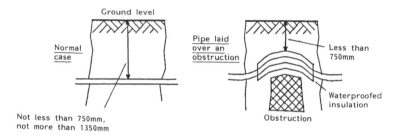

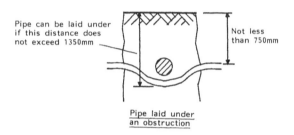

Diagram 92

BYELAW 49

PROTECTION FROM DAMAGE BY FREEZING AND OTHER CAUSES

> BYELAW 49. (1) Every pipe or other water fitting whether installed inside or outside a building or structure shall, so far as is reasonably practicable, be effectively protected -
>
> (a) whether by the manner of its installation, by insulation, or by some other no less effective means, against damage from freezing; and
>
> (b) against damage from other causes.

GENERAL GUIDANCE BYELAW 49

Generally installations will be accepted as satisfying the requirements of this byelaw if they were protected from freezing in accordance with BS 6700. The following table, based on that taken from this specification, gives recommended thicknesses of differing types of material.

Minimum thickness of thermal insulating material to delay freezing for frost protection:

NOMINAL OUTSIDE DIAMETER OF PIPE *	THERMAL CONDUCTIVITY W/(mk) OF INSULATING MATERIAL NOT EXCEEDING:							
	0.035 W/(m. K)	0.04 W/(m. K.)	0.055 W/(m. K.)	0.07 W/(m. K.)	0.035 W/(m. K.)	0.04 W/(m. K.)	0.055 W/(m. K.)	0.07 W/(m. K.)
	Indoor installations				Outdoor installations			
mm	mm	mm	mm	mm	mm	mm	mm	mm
Up to and including 15	22	32	50	89	27	38	63	100
Over 15 upto and including 22	22	32	50	75	27	38	63	100
Over 22 upto and including 42	22	32	50	75	27	38	63	89
Over 42 upto and including 54	16	25	44	63	19	32	50	75
Over 54 upto and including 76.1	13	25	32	50	16	25	44	63
Over 76.1 and flat Surfaces	13	19	25	38	16	25	32	50

NOTE   This table is based on discussions with the manufacturers pending revision of BS 5422. It lists the thermal conductivity value with an air temperature of 0°C and the minimum thickness of insulating material that will afford worthwhile protection against freezing during normal occupation of buildings.

* Watts per metre thickness per one degree Kelvin.

Examples of thermal conductivity of materials are:

Less than 0.035   - Polyurethane foam, foamed or expanded plastics including rigid and flexible preformed pipe insulation in these materials.

0.04 to 0.055   - Corkboard.

0.055 to 0.07   - Exfoliated vermiculite (loose fill).

It should be noted that smaller pipes require relatively greater thickness of material than large and indoor pipes up to 22mm nominal outside diameter require not less than 22mm finished thickness of mineral wool or polyurethane foam.

GUIDANCE BYELAW 49

This byelaw required pipes, fittings and appliances to be installed in positions which reduce risk of freezing. This especially applies to backflow prevention devices on which reliance is placed against contamination of drinking water. Any pipes, valves and fittings which cannot be so placed shall be effectively insulated against frost and supply pipes shall be capable of being drained. This applies in particular to pipes and fittings above ground and outside buildings and in which case insulation must be waterproofed. Generally pipes to outside taps and hose connections fitted with a servicing valve and draining valve inside a building, will be accepted as meeting the requirements of this byelaw if insulated as above.

For insulation requirements in ventilated roof spaces refer to BS 6700.

Location of pipes in buildings

Pipes shall not be installed without effective insulation in the following examples:

(a) unheated roof spaces,
(b) unheated cellars,
(c) unheated outbuildings including garages,
(d) near windows, air bricks, ventilators, external doors or where cold draughts are likely to occur,
(e) in any chase or duct formed in an external wall and in unheated ducts generally,
(f) in contact with any cold surface such as the internal face of an external wall.

Diagrams 93 to 97 show examples of acceptable ways of arranging pipes entering buildings.

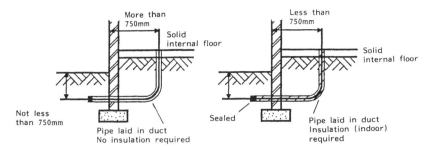

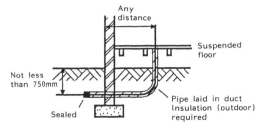

Diagram 93 - Pipes entering a building

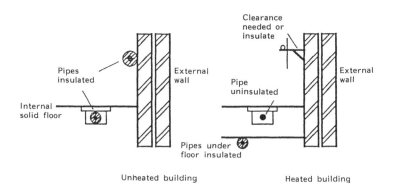

Diagram 94 - Pipes attached to external walls and laid in or under ground floors

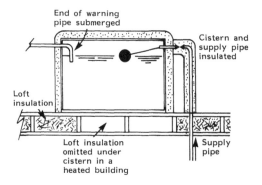

Diagram 95 - Cisterns in roof spaces

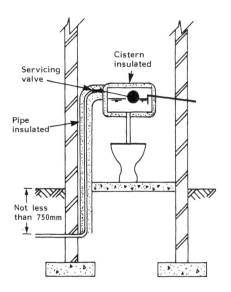

**Diagram 96 - Outside unheated
lavatories**

124

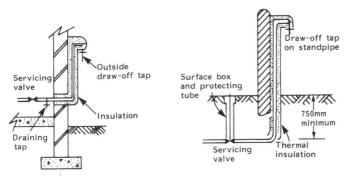

Diagram 97 - Examples of protection of outside taps and standpipes

(Backflow protection is not shown)

BYELAW 50

PROTECTION OF PLASTICS PIPES FROM PETROLEUM PRODUCTS

BYELAW 50. Every pipe made of plastics which is likely to be damaged by exposure to oil or petrol shall, so far as is reasonably practicable, be covered or otherwise effectively protected from such damage.

GUIDANCE BYELAW 50

Pipes of plastics must not be laid in ground subject to spillage of hydrocarbons such as oil, petrol or creosote which can cause deterioration of the pipe and consequent contamination and waste of water (see diagram 98). In addition to the case illustrated, taste problems have arisen where plastics pipes have been laid in close contact with damp proofing materials.

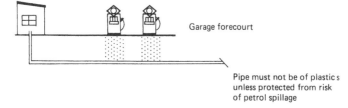

Garage forecourt

Pipe must not be of plastics unless protected from risk of petrol spillage

Diagram 98

PART VI

PREVENTION OF WASTE FROM OR CONTAMINATION BY UNSUITABLE OR IMPROPERLY INSTALLED WATER FITTINGS

CONTENTS

PART VI - PREVENTION OF WASTE FROM, OR CONTAMINATION BY UNSUITABLE OR IMPROPERLY INSTALLED WATER FITTINGS

BYELAWS 51 and 52

MATERIALS AND CONSTRUCTION OF WATER FITTINGS

BYELAW 51. No water fitting which conveys water supplied by the undertakers for domestic purposes shall -

> (a) be made wholly or partially of, or incorporate, or
>
> (b) be lined or coated with,

any material or substance which contaminates, or is likely to contaminate, such water by altering its colour, odour, taste or composition.

BYELAW 52. Every water fitting shall be constructed of materials, the nature, the strength and thickness of which (including any internal lining or external coating) will prevent, so far as is reasonably practicable, damage from -

> (a) any external load;
>
> (b) vibration, stress or settlement;
>
> (c) internal water pressure;
>
> (d) internal and external temperatures; and
>
> (e) corrosion.

GUIDANCE BYELAWS 51 AND 52 - PIPES, PIPE JOINTS AND PIPE FITTINGS

1. GENERAL

The requirements of byelaws 51 and 52 in respect of pipes and pipe fittings would be accepted as being satisfied if:

> (a) any pipe joint or pipe fitting complies with the specifications and additional requirements identified under the section of the particular

129

material given below. Special attention should be given to the distinction made between pipes laid below ground and those laid above. Should an installer contemplate using materials other than those mentioned below, he should consult the Directory and/or the undertakers before commencing work, and ensure that...

(b) it is installed in accordance with relevant recommendations of BS: CP 2010: Pipelines (currently under revision and is being reissued as BS 8010) and CP 312: Plastics pipework (thermoplastics material) and BS 6700, and

(c) it meets the requirements of Part II of these byelaws.

There may be special circumstances, e.g. when laying iron pipes in very acidic soils or concrete pipes in sulphated soils, where special additional precautions should be taken and the advice of the undertakers should be sought.

Note that no metal pipe joint laid in the ground shall be made using adhesives (see byelaw 57).

Some waters can cause a form of corrosion whereby a meringue-like product grows in or on certain brass fittings. This process of dezincification may not only choke the pipe or fitting but ultimately cause leakages and fittings to become inoperable. The requirements of byelaw 52(e) in respect of dezincification will be accepted as being satisfied if a water fitting or any part made of copper alloy containing zinc, if laid in the ground, is made of gunmetal or of a brass resistant to dezincification. All gunmetals and certain types of brass are immune and, recently, manufacturers have produced brasses suitable for hot stamping which are also resistant to dezincification. These meet the requirements for fittings to be laid in the ground and are usually marked by a nationally agreed symbol (see diagram 99). See also byelaw 53. It should be noted that in certain areas, water undertakers also restrict the use, above ground, of fittings which are not resistant or immune to dezincification.

Diagram 99

2. STEEL PIPES

(a) BS 1387: Screwed and socketed steel tubes and tubulars in association with BS 1740: Part 1 Wrought steel pipe fittings

Additional requirements:

If laid in the ground pipes, joints and fittings are to be:

i)   heavy gauge tube only, and

ii)   protected by the methods set out in BS 534: Section 30 External protection, and Section 31 to 33 Internal protection.

NOTE   Galvanising is not accepted as the only protection.

If installed above ground pipes, joints and fittings are to be:

i)   heavy or medium gauge tube except, where installed as part of a fire sprinkler system from which no water is drawn or used for other purposes, light gauge tube may be used, and

ii)   coated externally and internally in accordance with Sections 28 and 29 of BS 534 or galvanised in accordance with Section 4.3, and in some areas where water is aggressive to zinc additional internal protection precautions should be taken.   The undertakers' advice should be sought; and

iii)   if screwed joints, made using PTFE tape or paste as listed in the Directory.

NOTE   Pipes to BS 1387 are identified at each end by a ring of paint giving the class colour i.e.   light gauge is brown, medium gauge is blue, and heavy gauge is red.   Diagram 100 illustrates the application of each gauge.

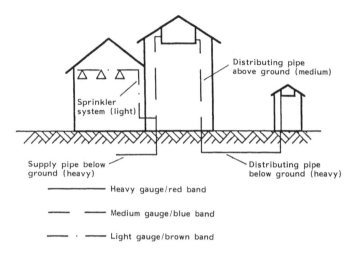

Diagram 100

Pipes are normally delivered in random lengths of between 4m and 7m but exact lengths can be supplied if required. All classes will withstand a works test pressure of 50 bar.

(b) BS 534: Steel pipes and specials for water and sewage

NOTE This specification requires that materials shall be to BS 3601 and dimensions selected from Table 1 of BS 3600.

Additional requirements:

If laid in the ground:

Pipes and fittings are to be protected by the methods set out in BS 534: Section 30 External protection and Sections 31 to 33 Internal protection.

If installed above ground:

Pipes and fittings are to be coated externally and internally in accordance with Sections 28 and 29 of BS 534.

(c) Acceptable pipe joints include:

BS 21:   Pipe threads for tubes and fittings where pressure tight joints are made on the threads (metric dimensions).

BS 143 and 1256:   Malleable cast iron and cast copper alloy threaded pipe fittings.

BS 1965:   Butt-welding pipe fittings for pressure purposes.   Part 1 Carbon steel.

BS 4504:   Flanges and bolting for pipes, valves and fittings, metric series.   Part 1 - Ferrous.

BS 2494:   Elastomeric joint rings for pipework and pipelines.

Subject to the protection specified for the pipes themselves.

3. IRON PIPES

BS 4622:   Grey iron pipes and fittings (production has now mainly ceased).

BS 4772:   Ductile iron pipes and fittings.

Additional requirements:

If laid in the ground:

Pipes and fittings are coated externally in accordance with Section 3.1 of BS 4772 and lined internally with cement mortar in accordance with Section 3.2 of the standard. In some areas the undertakers may require that, additionally, the pipes should be laid in a loose polyethylene sleeve to BS 6076. Grey iron pipe should be of the appropriate class for the working pressure and in no case less than Class 2.

If installed above ground:

Pipes and fittings are coated internally and externally in accordance with BS 4772: Sections 3.1 (and section 3.2 if cement mortar lining is required).

NOTES (i) Ductile iron pipes up to 300mm diameter are tested at works to 50 bar but should not be tested, when installed, to more than 45 bar. The complete installation should be considered as being suitable for working pressures of at least 12 bar.

(ii) Joints are of a proprietary nature but only flanged joints are covered by a British Standard. Examples are given below.

(a) Push in joint

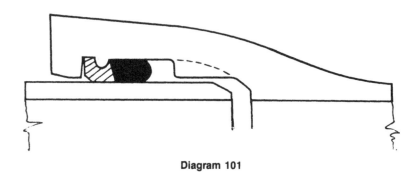

Diagram 101

This is a push in joint incorporating a specially shaped rubber gasket for pipelines conveying liquids. It can be deflected by several degrees in any direction and is capable of axial movement. It is used in ductile spun iron pipes and fittings in nominal sizes DN 80 - 600mm.

(b) Push in joint with self anchoring gasket

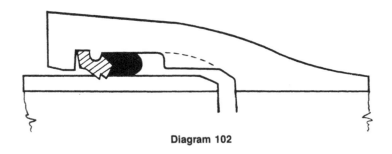

Diagram 102

The self anchor jointing system employs standard production ductile spun iron pipes and fittings in conjunction with a modified version of the gasket. The gasket is of standard design in respect of dimensions and shape but stainless steel toothed inserts are moulded into the gasket and grip into the pipe surface on attempted withdrawal.

(c) Typical mechanical joint

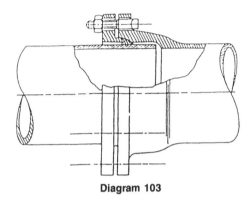

Diagram 103

The seal is effected by compressing a wedge shaped gasket on to the jointing surface in the pipe socket and the outside of the pipe spigot by means of a pressure gland and a series of bolts and nuts. The joint may be deflected to a limited degree in any direction and is capable of axial movement.

(d) Flanged joints BS 4504: Flanges and bolting for pipes, valves and fittings: Part 1 - Ferrous

Flanged joints are self anchoring and therefore external anchorages are not required at changes of direction and at blank ends. The joint is used mainly for "above ground" applications, e.g. in pumping stations and for industrial pipework. It is also commonly used to facilitate the installation and removal of valves in spigot and socket pipelines below ground and for valve bypass arrangements.

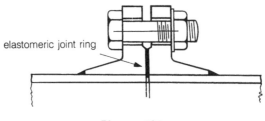

elastomeric joint ring

Diagram 104

Flanged ductile iron pipes are manufactured in standard lengths of 5m using either screwed or welded flanges.

(e) BS 2494: Elastomeric joint rings for pipework and pipelines

NOTE   For potable water the material used should be type W.

4. COPPER TUBES

BS 2871:   Copper and copper alloys, Tubes: Part 1 Copper tubes for water, gas and sanitation

Copper tube to this standard is designed to be jointed by compression or capillary fittings or by suitable methods of welding in some cases.

Copper tube shall either bear a BS certification mark or certification in a form acceptable to the undertakers which states that it has been effectively cleaned during, or subsequent to, manufacture.

Additional requirements:

If laid in the ground:

Copper tube must conform to Table Y.

In some areas the undertakers may advise that, additionally, copper tube is externally coated with a works applied polyethylene coating.

If installed above ground:

Copper tube to Table Z must not be bent or connected other than by capillary or non-manipulative type compression fittings.

136

Joints and fittings for use with BS 2871

BS 4504:   Flanges and bolting for pipes, valves and fittings. Metric series. Part 2 Copper alloy and composite flanges.

BS 1724:   Bronze welding by gas.

BS 864:   Capillary and compression tube fittings of copper and copper alloy Part 2 Specification for capillary and compression fittings for copper tubes (see diagrams 105 to 107).

Additional requirements:

If laid in the ground:

Compression fittings shall be of Type B to BS 864: Part 2.

Copper alloy fittings shall be made from gunmetal, BS 2872 or BS 2874 material CZ 132 (resistant to dezincification).

NOTE   that no pipe joint in the ground shall be made using adhesives.

If installed above ground:

The undertakers may also advise that, with certain exceptions, above ground fittings should be resistant to dezincification (see above).

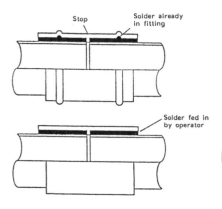

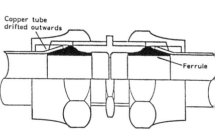

Diagram 105 - BS 864: Part 2
Capillary fittings for copper tube
suitable for any copper tube to BS
2871 for joints above or below ground.
Solders must be lead free.

Diagram 106 - BS 864: Part 2
Manipulative compression fittings
Type "B" Suitable for Tables X and Y
copper tube to BS 2871 for joints
above or below ground.

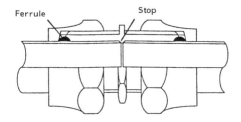

Diagram 107 - BS 864: Part 2 Non-manipulative compression fittings Type "A"
Suitable for use on any copper tube to BS 2871. For above ground use only.

5. ASBESTOS CEMENT PIPES

BS 486: Asbestos cement pressure pipes and joints

Additional requirements:

If laid in the ground:

Asbestos cement pipes and fittings are tested to a minimum works hydraulic test pressure of 15 bar or twice the maximum working pressure, whichever is the higher.

In some areas the characteristics of the water supply or the aggressive nature of the soil make it necessary to protect asbestos cement pipes and the advice of the undertakers should be sought.

If installed above ground:

Pipes and fittings are of the appropriate class to withstand a minimum test pressure 1.5 times the maximum working pressure.

CLASS	WORKS TEST PRESSURE (bar)(	MAXIMUM ALLOWABLE WORKING PRESSURE (bar)	METRIC COLOUR CODING (PIPES, JOINTS AND RINGS)
15	15	7.5	Green
20	20	10.0	Blue
25	25	12.5	Violet
Where a particular joint or rubber ring is common to two classes of pipe the colour coding relates to the higher class			

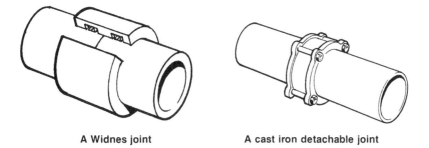

A Widnes joint   A cast iron detachable joint

Diagram 108 - Two joints for asbestos cement pipelines

The cast iron detachable joint sometimes referred to as the "Gibault" joint has an extended centre collar and two flanges. Two round rubber rings provide the sealing medium, the joint being compressed by means of specially treated bolts. Made in sizes 50mm to 250mm diameter.

6. UNPLASTICISED PVC PIPES

BS 3505: Unplasticised polyvinyl chloride (PVC-U) pressure pipes for cold potable water

This pipe conforms to imperial sizes.

Additional requirements:

If laid in the ground:

PVC-U pipe shall be of the appropriate class for the maximum working pressure but not less than 12 bar for sizes up to and including size 2.

Pipes of nominal size below 3 are not recommended for service installations.

BS 4346: Joints and fittings for use with unplasticised PVC pressure pipes

Part 1 Injection moulded unplasticised PVC fittings for solvent welding for use with pressure pipes, including potable water supply.

Part 2 Mechanical joints and fittings principally of unplasticised PVC.

Part 3 Solvent cement.

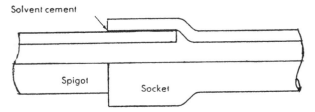

uPVC pipe to BS 3505 with solvent cement joint

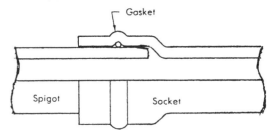

uPVC pipe to BS 3505 with mechanical joint

Diagram 109 - Joints for uPVC pipes

NOTE   Compression fittings similar to those used for polyethylene pipes may also be used on PVC-U. If used on pipes laid below ground they must be made of dezincification resistant or immune material.

7. POLYETHYLENE PIPES

BS 6572:   **Blue polyethylene pipes up to nominal size 63 for below ground use for potable water.** This pipe conforms to metric series dimensions.

Additional requirements:

This pipe is specially developed for services laid in the ground. It is pigmented blue and there is one class designed for a working pressure of 12 bar at 20C in sizes up to nominal size 63. It may be used above ground in situations where it is not exposed to direct sunlight.

Pipes larger than nominal size 63 are covered at present in nominal sizes 90 to 1 000 by WAA-SWMC Information and Guidance Note No. 4-32-03 (ISSN 0267-0305).

BS 6730:   **Black polyethylene pipes up to nominal size 63 for above ground use for cold potable water**

This pipe conforms to metric series dimensions and has been specially developed for above ground use for optimum resistance to ultra-violet light in situations where pipes are exposed to direct sunlight. N.B. Pipes larger than nominal size 63 will be covered by IGN No. 4-32-09 in due course.

BS 864: Part 5 Compression fittings of copper and copper alloy for polyethylene pipes with outside diameters to BS 5556 (Standard under preparation and will replace WAA-SWMC Information and Guidance Note No. 4-22-01 ISSN 0267-0305).

These fittings conform to metric series dimensions and are suitable for use with BS 6572 (blue) or BS 6730 (black) polyethylene pipe.

Additional requirements:

If laid in the ground:

It is essential that the method of jointing and the material used in the joint are suitable for use underground.

Copper alloy fittings made from gunmetal or material to BS 2872 or BS 2874 CZ 132 are acceptable in meeting the requirement to be dezincification resistant.

WAA-SWMC IGN 4-32-04 ISSN 0267-0305 Polyethylene socket and spigot joints and fittings, saddles and drawn bends for fusion jointing for use with cold potable water polyethylene pressure pipes.

This covers fusion fittings for pipes with outside diameters to BS 5556 (metric) in the nominal size range 20 to 1000, in both blue and black pigmented polyethylene, for use in cold potable water services at pressures up to 10 bar or 12 bar at 20C according to size. The IGN is under review and a new version is to be published.

A specification for electrofusion fittings, IGN No. 4-32-06 is in the course of preparation.

PE pipe fusion welding should be performed in accordance with WAA-SWMC IGN No. 4-32-08 Specification for site fusion jointing of MDPE pipe and fittings.

BS 1972: Polythene pipe (Type 32) for above ground use for cold water services.

This pipe is coloured black for above ground use only and conforms to imperial sizes. The standard has effectively been replaced by BS 6730 and may be withdrawn.

8. BS 4991: PROPYLENE COPOLYMER PRESSURE PIPE (SERIES 1)

This pipe in series 1 form conforms to imperial size dimensions and should only be used where the normal sustained working temperature does not exceed 20C.

Additional requirements:

If laid in the ground:

Should be designed for a working pressure of 12 bar and to avoid confusion with other services it must be identifiable by the colour blue.

If installed above ground:

Should be of the appropriate class to withstand a minimum test pressure of either twice or 1.5 times the maximum working pressure depending on its location (see byelaw 53).

9. BS 864: PART 3 COMPRESSION FITTINGS FOR POLYETHYLENE PIPES

These fittings conform to imperial size dimensions and are suitable for use with pipes to BS 1972 and BS 4991. The performance requirements are set out in BS 5114.

Additional requirements:

If laid in the ground:

The method of jointing and the material used for the joint must be suitable. Copper alloy fittings made from gunmetal or material to BS 2872, or BS 2874 CZ 132 are acceptable in meeting the requirement to be dezincification resistant.

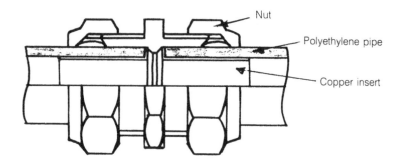

Nut

Polyethylene pipe

Copper insert

Diagram 110 - Typical joint for use on polyethylene (and PVC-U pipe)

OTHER SPECIFICATIONS FOR POLYETHYLENE PIPE JOINTS

Fittings other than copper and copper alloy fittings to BS 864: Parts 3 and 4 are available and those that are listed in the Directory are, conditionally or unconditionally, acceptable.

10. STAINLESS STEEL PIPES

BS 4127: **Light gauge stainless steel tubes: Part 2 Metric units.** This tube is suitable for use with BS 864: Part 2 Capillary or compression fittings.

Additional requirements:

If laid in the ground:

The class of pipe should be of the appropriate class for the maximum working pressure but not less than 12 bar in diameters up to and including 63mm. Joints made using adhesives are not permitted.

If installed above ground:

Pipes and fittings should be of the appropriate class to withstand a minimum test pressure 1.5 times the maximum working pressure.

11. MATERIALS AND CONSTRUCTION OF CISTERNS, TANKS AND CYLINDERS

The requirements of byelaws 51 and 52 in respect of cisterns, tanks and cylinders would be accepted as being satisfied if in the case of any new cistern, tank or cylinder:

(a) materials in contact with water meet the requirements of byelaws 7(1) and 8, and

(b) the cistern is provided with a rigid cover, and

(c) it is of the appropriate class or grade, and

(d) it complies with any of the following standards:

* BS 417: Galvanised low carbon steel cisterns, cistern lids, tanks and cylinders. Part 2 Metric units.

** BS 699: Copper direct cylinders for domestic purposes.

BS 843: Thermal storage electric water heaters (constructional and water requirements).

** BS 853: Calorifiers and storage vessels for central heating and hot water supply.

BS 1125: WC flushing cisterns (including dual flush cisterns and flush pipes).

* BS 1563: Cast iron sectional tanks (rectangular)(Subject to being able to pass watertightness tests on site).

* BS 1564: Pressed steel sectional rectangular tanks (Subject to being able to pass watertightness test on site).

* BS 1565: Galvanised mild steel indirect cylinders, annular or saddle back type. Part 2 Metric units.

** BS 1566: Copper indirect cylinders for domestic purposes. Part 1 Double feed indirect cylinders.

** BS 1566: Copper indirect cylinders for domestic purposes. Part 2 Single feed indirect cylinders.

BS 1876: Automatic flushing cisterns for urinals.

145

BS 2594: Carbon steel welded horizontal cylindrical storage tanks.

** BS 3198: Copper hot water storage combination units for domestic purposes.

BS 4213: Cold water storage and feed and expansion cisterns, (polyolefin or olefin copolymer) and cistern lids.

NOTE * Subject to being protected against corrosion, where relevant, in accordance with BS 3416 (Type II material only).

** Subject to being protected against corrosion, where relevant, with sacrificial anodes made from alloy 1050A to BS 1475.

BYELAWS 53 and 54

LAYING AND JOINTING OF PIPES

BYELAW 53. Every water fitting which -

    (a) is installed below ground; or

    (b) passes through or under any wall, footing or foundation; or

    (c) is embedded in any wall or solid floors; or

    (d) is enclosed in any chase or duct; or

    (e) is in any other position which is inaccessible, or to which access is difficult;

shall be -

    (i) constructed to withstand without bursting, buckling, fracture or leaking an internal hydraulic pressure twice that to which it would normally be subject; and

    (ii) installed to accommodate any reasonably foreseeable movement (including any thermal movement) in the pipe; and

    (iii) except in a closed circuit, resistant to dezincification.

GUIDANCE BYELAWS 53 AND 54 - LAYING AND JOINTING OF
PIPES

General

The requirements of byelaws 53 and 54 will be accepted as being
satisfied if pipes and fittings are installed in accordance with BS 6700
or CP 312 where relevant.

Anchoring pipes

Pipes laid below ground or beneath floors, etc.   must be capable of
meeting a test pressure of twice the maximum working pressure and
care should be taken to secure and anchor pipes adequately.   The
following diagrams illustrate acceptable means of restraining pipes
against thrust at bends, tees, etc.

NOTE   for further information about thrust blocks see BS 6700.

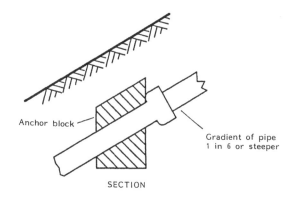

SECTION

Diagram 111 - Gradient thrust blocks (buried or exposed mains)

Gradient	Spacing of Anchor block	
1 in 2	18ft	5.5m
1 in 3	36ft	11.0m
1 in 4	36ft	11.0m
1 in 5	54ft	16.5m
1 in 6	72ft	22.00m

Diagram 111 (Cont) - Spacing of anchor blocks

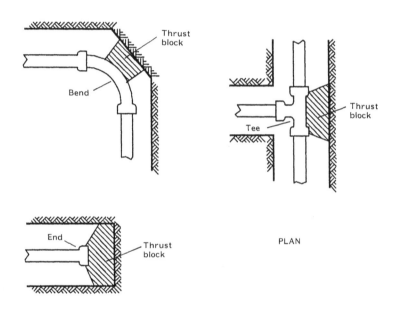

Diagram 112 - Horizontal thrust blocks

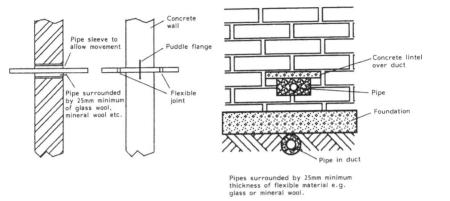

Diagram 113 - Examples of alternative ways of laying pipes through or under foundations underground (byelaw 53)

Expansion of pipes carrying hot water

The requirements of byelaw 53 would be accepted as being satisfied whenever expansion of pipes carrying hot water is catered for in adequate expansion bellows or expansion loops (see diagram 114).

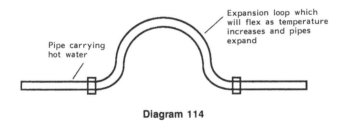

Diagram 114

GUIDANCE BYELAW 54

Air locks and water hammer

The requirements of byelaw 54 will be accepted as being satisfied in respect of removal of air locks if an installation has been designed to facilitate the removal of air during filling and pipes have a slight rise to a cistern, vent pipe or an automatic air release valve and should, wherever possible, fall to draw-off points.

Whilst there is no documentary evidence to show the magnitude and extent of waste of water in consumers' water services as a consequence of water hammer, there is no doubt that sudden excessive pressure rises within pipework could lead to premature failure of jointing and possible damage to connected fittings. In cases where unacceptable water hammer occurs when particular fittings or appliances are used, suitable measures should be taken which could include the fitting of air or gas loaded vessels or special mechanical water hammer arresters.

BYELAW 55

CLEANING PIPES AFTER INSTALLATION OR REPAIR

BYELAW 55. Every pipe which supplies or may supply, water for domestic purposes shall be flushed to remove debris before it is first used after installation, renewal or repair.

GUIDANCE BYELAW 55

Before taking into service flushing is necessary to remove debris such as filings, wire wool, flux residues, etc. It should be carried out systematically so that debris is not washed into a section of pipework that has already been cleansed. Attention is drawn to relevant recommendations in BS 6700.

BYELAW 56

PREVENTION OF WARMING OF WATER IN PIPES

BYELAW 56. Every supply pipe installed in a dwelling which supplies cold water for domestic purposes to any tap, shall be installed in such a place or position that, so far as is reasonably practicable, the water will not be warm when it is drawn off from that tap.

GUIDANCE BYELAW 56

The requirements of byelaw 56 would be accepted as being satisfied if the layout of cold water pipes and cisterns keeps them remote from sources of heat, hot water pipes and warm spaces generally.

In circumstances where a cold water pipe cannot be located except in close proximity to a permanent source of heat or in a heated environment, e.g. in an airing cupboard, in ducts, etc. the pipe must be adequately insulated. The connection of cold water draw-off taps or fittings specifically designed to cool water is not an acceptable means of satisfying this byelaw.

BYELAW 57

USE OF ADHESIVES IN JOINTING METAL PIPES

> BYELAW 57. No metal pipe which -
>
> > (a) is installed in the ground or passes through or under any wall, footing or foundation; or
> >
> > (b) is embedded in any wall or solid floor; or
> >
> > (c) is enclosed in any chase or duct; or
> >
> > (d) is in any other place or position to which access is difficult;
>
> shall be connected to any other water fitting by means of any adhesive.

GUIDANCE BYELAW 57

This byelaw would prohibit the jointing of stainless steel pipes below ground or other inaccessible position by the use of adhesives.

BYELAW 58

ACCESSIBILITY OF PIPES AND PIPE FITTINGS

> BYELAW 58. (1) No pipe or other water fitting shall be embedded in any wall or solid floor or installed in or below a solid floor or under a suspended floor at ground level.
>
> (2) Paragraph (1) shall not apply to -
>
> > (a) a pipe or other water fitting installed in a chase or duct (not being the cavity in a cavity wall) in a wall or solid floor which may, if necessary, be readily exposed; or

151

GUIDANCE BYELAW 58

Detailed guidance cannot be given for every circumstance likely to be encountered. A number of illustrative examples are given in diagram 115. These are based on the principles:

(a) That enclosures within chases and ducts can be permitted so long as leaks would become apparent and the section of pipe could be exposed by the removal of covers or superficial finishes or the pipe could be withdrawn for repair.

(b) More stringent precautions should be taken with pipes below ground floor level and other situations where leaks could be undetected for long periods.

(c) Particular care should be taken where continuous flooring such as chipboard is laid. Properly formed openings with removable covers should be provided to give adequate access for inspection and dismantling joints and for removal of sections of pipe.

(d) Where insulation is required this should be in accordance with the recommendations of BS 6700 (see comment on byelaw 49).

NOTE   The bedding of any pipe and associated pipe joints forming part of a closed circuit system of underfloor space heating in screed or in a properly formed chase in a wall or solid floor which is subsequently plastered or screeded will be accepted if the pipe and joints can be exposed for repair or replacement by removing the surface layers of plaster or screed.

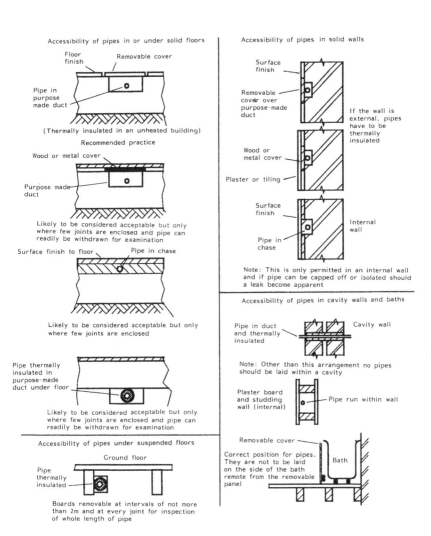

Accessibility of pipes in or under solid floors

Floor finish   Removable cover

Pipe in purpose made duct

(Thermally insulated in an unheated building)

Recommended practice

Wood or metal cover

Purpose made duct

Likely to be considered acceptable but only where few joints are enclosed and pipe can readily be withdrawn for examination

Surface finish to floor   Pipe in chase

Likely to be considered acceptable but only where few joints are enclosed

Pipe thermally insulated in purpose-made duct under floor

Likely to be considered acceptable but only where few joints are enclosed and pipe can readily be withdrawn for examination

Accessibility of pipes under suspended floors

Ground floor

Pipe thermally insulated

Boards removable at intervals of not more than 2m and at every joint for inspection of whole length of pipe

Accessibility of pipes in solid walls

Surface finish

Removable cover over purpose-made duct

If the wall is external, pipes have to be thermally insulated

Wood or metal cover

Plaster or tiling

Surface finish

Pipe in chase

Internal wall

Note: This is only permitted in an internal wall and if pipe can be capped off or isolated should a leak become apparent

Accessibility of pipes in cavity walls and baths

Pipe in duct and thermally insulated   Cavity wall

Note: Other than this arrangement no pipes should be laid within a cavity

Plaster board and studding wall (internal)   Pipe run within wall

Removable cover

Correct position for pipes. They are not to be laid on the side of the bath remote from the removable panel   Bath

Diagram 115

153

BYELAW 59

DISCONNECTION OF WATER FITTINGS

> **BYELAW 59.** (1) Every pipe, or part of a pipe which conveys water to a water fitting shall, if that water fitting is disconnected, be disconnected.
>
> (2) Paragraph (1) shall not apply where a water fitting is disconnected for repair or renewal and is replaced within a period of 60 days.

GUIDANCE BYELAW 59

This byelaw is to prevent contamination due to water in any unused pipework becoming stagnant.

Assuming that the house shown in diagram 116 has been modernised and that the outside WC will no longer be required it would not be permissible to seal off the supply pipe at point B. The disconnection would have to be made at point A.

Note also that it is not acceptable to lay branch pipes to stop ends to facilitate connection of future appliances unless there is a physical break at the junction with the supply pipe.

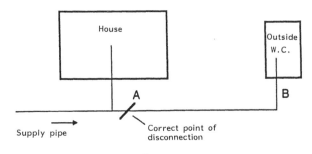

Diagram 116

NOTE A fitting or appliance may remain unconnected for a period not exceeding 60 days to allow a reasonable amount of time for installers to obtain replacements for broken or damaged fittings or appliances.

BYELAW 60

CONNECTING WATER FITTINGS OF DISSIMILAR METALS

BYELAW 60. No metal pipe or pipe joint or other water fitting shall be connected to any other pipe, pipe joint or other water fitting constructed of a different metal (whether or not by way of repair or replacement) unless either -

(a) deterioration through galvanic action is unlikely to occur; or

(b) effective measures are taken to prevent such deterioration.

Pipes and fittings of dissimilar metals

This byelaw is to protect against two forms of galvanic action which can cause severe corrosion of pipes. Direct galvanic action takes place where two dissimilar metals are in contact. Indirect galvanic action is where dissolution of one metal upstream of another causes accelerated corrosion of the latter. The characteristics of the water have an important bearing on the severity or otherwise of the action.

The requirements of the byelaw would be accepted as being satisfied if the sequence of metals in any supply or distributing pipes including connected cisterns or cylinders in relation to the normal direction of flow is:

Direction of flow	Galvanised Steel
	Uncoated iron
	Lead
	Copper

Care should be taken in the use of metal jointing materials which should in general conform with the above sequence.

Plastics pipes may be used in conjunction with any material. Although copper pipe should not be used in the replacement of a section of lead pipe, the use of one or two short copper liners in repairs should not result in water picking up a significant concentration of copper.

Problems can arise in cases where copper flow and return pipes are connected to galvanised hot water tanks or to galvanised storage cisterns used in conjunction with copper cylinders.

155

A simple explanation of what occurs when dissimilar metals are joined or connected together is that an electric cell is set up across the joint. This is basically the same as a wet cell battery except that in this case the water acts as the electrolyte (i.e. a liquid capable of carrying electricity). The resultant current flow, though minute, causes one metal to corrode, and eventually perforation occurs.

Another form of galvanic action is that in which particles of one metal are taken into solution and the resultant solution attacks a pipe or fitting of a different metal further along the pipework system. An example of this is when the water, having passed through copper pipe to a galvanised cistern, then stands in the cistern for some time and an attack on the galvanising takes place: eventually perforation will occur.

Even though the use of an intervening non-conductive material will prevent direct galvanic action at the joint of two different metals, this will not prevent the occurrence of the latter type of corrosion. The position of the metals in the pipework system then becomes most important. The obvious answer is to use only one metal in the system but as this is often impracticable, precautions can be taken by using non-conductive materials, e.g. polyethylene, etc.

In some circumstances cathodic protection can give protection against galvanic action. This involves using sacrificial anodes which are fitted into cisterns, cylinders or tanks and on pipelines so that the anode will corrode instead of the fitting it protects until a protective coating is formed on the latter. Fittings made from galvanised steel are frequently protected in this way.

In the case of galvanised metal the anodes are usually made from magnesium alloy, fitted on to steel rods and fixed in the middle of the fittings so that they are immersed in the water. In the case of copper cylinders and calorifiers anodes are usually made of aluminium rod (see diagram 117).

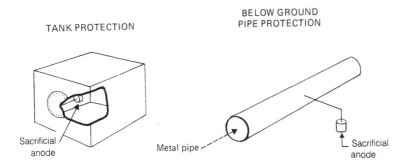

TANK PROTECTION

BELOW GROUND
PIPE PROTECTION

Sacrificial
anode

Metal pipe

Sacrificial
anode

Galvanised steel tank　　　　　**Steel or iron pipes**

Diagram 117 - Examples of providing cathodic protection against galvanic corrosion by using sacrificial anodes

PART VI

STOP VALVES, ETC

CONTENTS

PART VII - STOPVALVES ETC.

BYELAWS 61, 62 and 63

PROVISION AND LOCATION OF STOPVALVES

BYELAW 61. In this Part of these byelaws "premises" means

    (a) any premises to which a separately chargeable supply of water is provided by the undertakers; and

    (b) any premises which are occupied as a dwelling whether or not separately charged for a supply of water.

BYELAW 62. (1) Every supply and distributing pipe providing water to premises shall be fitted with a stopvalve to enable the supply to those premises to be shut off without shutting off the supply to any other premises.

    (2) Every stopvalve mentioned in paragraph (1) shall, so far as is reasonably practicable, be -

    (a) inside premises;

    (b) above floor level;

    (c) as near as possible to the point where the supply first enters the premises; and

    (d) so installed that its closure will prevent the supply of water to any point of use.

BYELAW 63. (1) Every supply and distributing pipe providing water in common to two or more premises shall be fitted with a stopvalve (whether inside or outside premises) to which each occupier of premises has access.

    (2) Every stopvalve mentioned in paragraph (1) shall be installed so that its closure will prevent the supply of water to all of the premises supplied by that common pipe.

GUIDANCE BYELAWS 61, 62 AND 63

These byelaws detail stopvalves that must be provided in installations. They do not preclude fitting other "non statutory" stopvalves. The Water Acts normally require each premises to have a separate supply pipe wherever possible. The principle on which these byelaws is based is that if any occupiers are suffering damage or nuisance to their property due to a leaking or defective fitting whether on their own or on a common pipe they should have access to a stopvalve which controls the supply to that fitting.

The following diagrams (118 to 125) give examples of compliance with these byelaws. The possible additional need for secondary backflow protection should be noted (see guidance to byelaw 26).

Examples of installations complying with byelaws 62 and 63

Case (a) Premises to which a separately chargeable supply is provided.
Case (b) Premises occupied as a dwelling.

Case (a) Premises not separately chargeable
Case (b) B is occupied as a dwelling. A is not, e.g. house with an outbuilding.

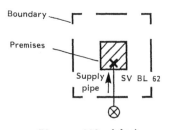

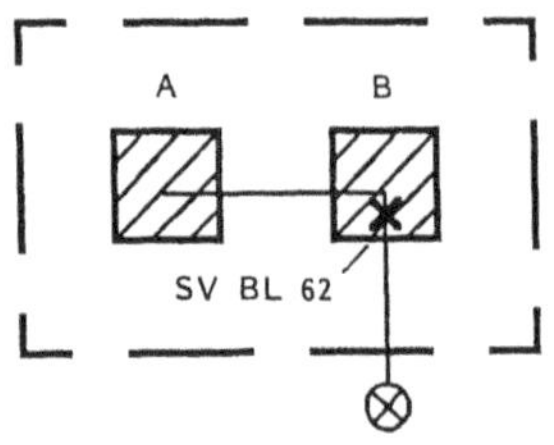

Diagram 118 - (plan)

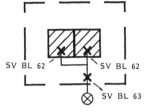

Diagram 119 - (Plan)

Case (a) Two separately chargeable premises on a common supply pipe.
Case (b) Two dwellings on a common supply pipe.

Case (a) Two separately chargeable parts of a building
Case (b) Two parts of a building, each occupied as a dwelling, e.g. a pair of semi-detached houses.

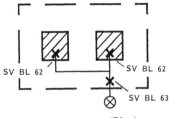

Diagram 120 - (Plan)

Diagram 121- (Plan)

162

Installations complying with byelaws 62 and 63 (cont)

Building with a number of supply pipes entering.
Case (a) parts separately chargeable,
Case (b) parts separately occupied as dwellings.
Case (c) a mixture of (a) and (b).

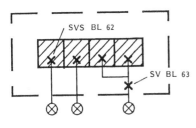

Diagram 122 - (Plan)

Building comprising several occupied parts (e.g. flats)each with separate supply pipes outside those parts.

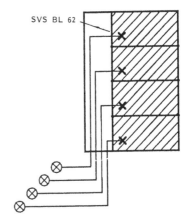

Diagram 123 - (Elevation)

Building comprising several separately occupied dwellings (e.g. flats) on a common supply pipe outside those parts.

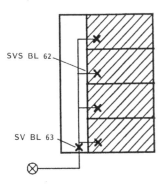

Diagram 124 - (Elevation)

Building comprising several separately occupied dwellings (e.g. flats) with a common distributing pipe outside those parts.

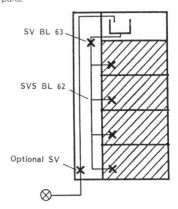

Diagram 125 - (Elevation)

163

Examples of installations which are either not permitted or not recommended. The advice of the undertakers should be sought before commencing any work.

Separately occupied or separately chargeable premises on a common supply pipe passing through one of those premises.

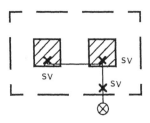

Diagram 126 - (Plan)

Building comprising several separately occupied or separately chargeable parts on a common supply pipe passing through some of those parts.

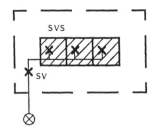

Diagram 127 - (Plan)

Building comprising two or more parts separately occupied or separately chargeable. Separate supply pipes passing throught some of those parts.

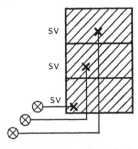

Diagram 128 - (Elevation)

Building comprising several separately occupied or separately chargeable parts. Common supply pipe passing through some of those parts.

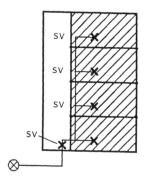

Diagram 129 - (Elevation)

Building comprising several parts either separately occupied or separately chargeable. Common distributing pipes passing through some of those parts.

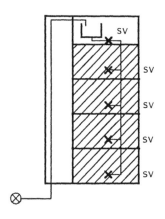

Diagram 130 - (Elevation)

Byelaw 62(2), which is to ensure that supply stopvalves fitted in compliance with byelaw 62(1) shall be above floor level and close to the point of entry, would be deemed to be satisfied if every stopvalve is within 150mm of the floor or wall through which the pipe enters the building or part of a building.

The byelaw also requires that a supply stopvalve should be fitted so that it controls the supply to any part of the installation in the building supplied, upstream of any branch pipe (see diagram 131).

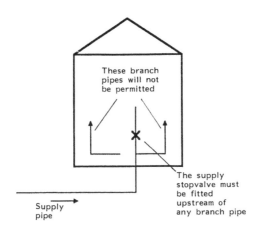

These branch pipes will not be permitted

The supply stopvalve must be fitted upstream of any branch pipe

Supply pipe

Diagram 131

BYELAW 64

REQUIREMENTS FOR STOPVALVES

BYELAW 64. Every stopvalve fitted in accordance with byelaws 62 and 63 shall -

(a) be watertight when closed; and

(b) be watertight when open and subjected to an internal hydraulic pressure 1.5 times the pressure to which it is normally subject; and

(c) except in the case of a plug valve, be so designed or adapted that its seal can be readily renewed; and

(d) not incorporate a loose washer plate; and

(e) be reasonably resistant to corrosion.

GUIDANCE BYELAW 64

Byelaw 64 will be deemed to be satisfied if every supply stopvalve is of an appropriate pressure rating for the supply pipe on which it is fitted and it complies with one of the following British Standard Specifications or is listed in the Directory.

Stopvalves located outside buildings

(a) BS 5433: Underground stopvalves for water services.

BS 5163: Predominantly key-operated cast iron gate valves for waterworks purposes.

BS 2580: Underground plug cocks for cold water services.

NOTE   BS 1010 stopvalves would be accepted outside buildings if installed above ground (see under (b) below).

Stopvalves located inside buildings

(b) as (a) above and

BS 1010: Draw off taps and stopvalves for water services (screw down pattern). Part 2 Draw off taps and above ground stop valves.

166

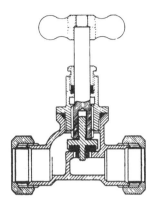

BS 5433 Underground stopvalve
BS 1010 Above ground stopvalve

BS 5163 Double flanged gate valve

BS 2580 Plugcock

Diagram 132 - Examples of stopvalves

BYELAWS 65, 66 and 67

DRAINING OF SUPPLY PIPE, REQUIREMENTS FOR AND
LOCATION OF DRAINING TAPS

BYELAW 65. Every supply pipe in premises shall be installed so
that when the stopvalve installed in accordance with byelaw 62 in
respect of those premises is closed, and any draining tap is open,
that pipe may be drained

BYELAW 66. Every supply pipe in premises shall be fitted with a
draining tap which -

(a) is watertight when closed and subjected to
an internal hydraulic pressure 1.5 times the
pressure to which it is normally subject;

(b) is so designed or adapted that its seal can
be readily renewed; and

(c) is reasonably resistant to corrosion.

BYELAW 67. No draining tap fitted to a supply pipe shall be -

(a) buried in or covered with soil;

(b) installed so that it is submerged, or is likely
to be submerged.

GUIDANCE BYELAWS 65 and 66

Byelaw 65 sets out the basic requirement that a supply pipe must be
capable of being drained. The requirements of byelaw 66 would be
satisfied if any draining tap complies with BS 2879: Draining taps (screw
down pattern), or with the relevant requirements of BS 1010: Part 2.

The definition of supply pipe includes any pipework which is subject to
water pressure from the undertakers' mains (whether or not any
pressure control device is employed). Therefore pipework downstream
of mains fed heaters, water softeners, etc. shall be capable of being
drained.

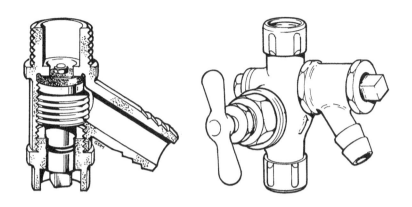

Diagram 133 - Draining Tap with "O"
ring gland seal BS 2879

Diagram 134 - BS 1010: Part 2
Combination stopvalve and BS 2879
Draining tap

NOTE   Types of draining taps to editions of BS 2879 prior to 1980 are not acceptable.

GUIDANCE BYELAW 67

Byelaw 67 is intended to eliminate the risk of backflow of contaminated water during a draining operation.   An example is given in diagram 135 where a draining tap is situated in a sump which may be flooded.   This would be prohibited.

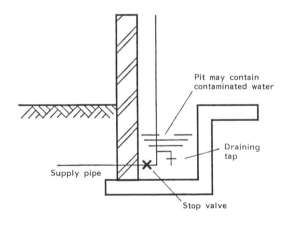

Diagram 135 - Example of prohibited draining tap

BYELAWS 68, 69 and 70

PROVISION AND LOCATION OF SERVICING VALVES

BYELAW 68.   Every pipe for conveying water from -

(a) a cold water storage cistern the capacity of which exceeds 18 litres; or

(b) a hot water storage cistern, cylinder or tank;

shall be fitted with a servicing valve as close to that cistern, cylinder or tank as is reasonably practicable.

BYELAW 69.   (1) Every pipe supplying water to a float-operated valve shall be fitted with a servicing valve to shut off the supply of water to that valve.

(2) Every servicing valve installed in accordance with paragraph (1) shall be fitted as near as is reasonably practicable to the float-operated valve.

BYELAW 70.   No servicing valve shall be installed as the sole means of controlling or preventing the flow of water through any pipe unless that valve complies with byelaw 64.

GUIDANCE BYELAW 68

Servicing valves are required on outlets from cisterns above 18 litres capacity so that if it is required to drain down the installation loss of water would be minimal.

The requirements of the byelaws will be accepted as being satisfied if any servicing valve is provided and located on a distributing pipe as shown in diagram 136.

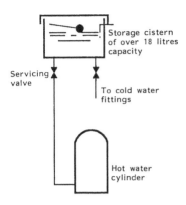

Diagram 136 - Separate cistern and hot water cylinder

In cases where hot water apparatus is arranged in such a way that it is not possible to control the pipe from the storage cistern, then it is permissible to fit a valve on every pipe as it leaves the hot water cylinder or tank. This applies to hot water combination units where the cold feed from the cistern to the hot water compartment is installed within the unit.

No servicing valve shall be fitted in any vent pipe.

GUIDANCE BYELAW 69

Byelaw 69 requires also that a servicing valve is fitted as near as practicable to every float-operated valve. This is to enable the supply to be shut off in the event of a faulty float or valve and for repairs to be effected without shutting off the supply to the rest of the installation. Examples are given in diagram 137. See also guidance to byelaw 49 concerning provision of servicing valves to outside taps and hose connections.

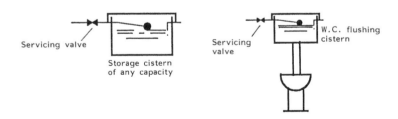

Diagram 137

171

GUIDANCE BYELAW 70

Byelaw 70 is intended to prevent servicing valves from being used as flow control or stop valves unless they meet the requirements for stop valves given above.

BYELAW 71

REQUIREMENTS FOR SERVICING VALVES

BYELAW 71. Every servicing valve shall be -

(a) designed or adapted to operate only by the insertion of a screwdriver or other instrument in a slot on the valve, except where it -

(i) complies with byelaw 64; or

(ii) is installed on a distributing pipe with a static water pressure not exceeding 1 bar; and

(b) watertight when closed; and

(c) capable of withstanding without leaking an internal hydraulic pressure 1.5 times the pressure to which is is ordinarily subject.

GUIDANCE BYELAW 71

The requirements of byelaw 71 will be accepted as being satisfied if any servicing valve, having regard to the working pressure and temperature to which it could be subjected, is of the appropriate rating and it complies with one of the standards identified in guidance to byelaw 64 or with BS 6675: Servicing valves (copper alloy) for water services.

The purpose of a distinction in the type of actuation of servicing valves is to ensure that, at least on supply pipes, they are unlikely to be used as control valves or be inadvertently operated.

For valves which comply refer to the Directory. These include certain lever operated ball type valves which are accepted as meeting the requirements of byelaw 71(a)(i).

BYELAW 72

WATERTIGHTNESS OF BACKFLOW PREVENTION DEVICES

BYELAW 72. Every vacuum breaker, check valve, double check valve assembly or combination of check valves and vacuum breakers installed in any pipe shall be -

    (a) watertight when closed; and

    (b) capable of withstanding without leaking an internal hydraulic pressure 1.5 times the pressure to which it is normally subject.

GUIDANCE BYELAW 72

The requirements of byelaw 72 insofar as watertightness is concerned will be accepted as being satisfied where:

    (a) any vacuum breaker complies with BS 6282: Part 3;

    (b) any check valve complies with BS 6282: Part 1, and

    (c) any combined check valve and vacuum breaker complies with BS 6282: Part 4.

In the case of vacuum breakers attention is drawn to the advice given by the UK Water Fittings Byelaws Scheme in the Directory:

" The installation must be such that any (operational) discharge can be readily observed and it is recommended that these valves be installed so that no consequential damage from any discharge can take place".

NOTE See guidance to byelaw 11 for details of these devices and their positioning.

BYELAW 73

ACCESSIBILITY OF STOPVALVES AND SERVICING VALVES

BYELAW 73. Every stopvalve and servicing valve installed in accordance with this Part of these byelaws shall be so placed that so far as is reasonably practicable it can be readily examined, maintained and operated.

GUIDANCE BYELAW 73

The key phrase in this byelaw is "readily examined, maintained and operated". Whether it be a stopvalve or a servicing valve it must be placed so that it can be reached without difficulty for turning on or off. Examples of compliance are given in diagram 138.

The byelaw requires that not only should valves be accessible for operating but access should be sufficient for maintenance purposes. This precludes burying valves in walls, floors, etc.

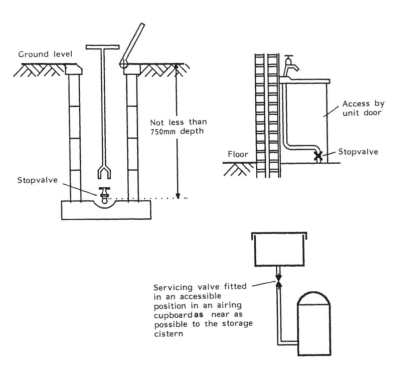

Diagram 138 - Examples of locations of valves complying with byelaw 73

PART VIII

WATERCLOSETS AND URINALS

CONTENTS

PART VIII - WATERCLOSETS AND URINALS

INTRODUCTION

The following diagrams illustrate some of the terms used in guidance to Part VIII:

Flushing cistern

When lever in diagram 139 is depressed the plunger rises pushing air out of the siphon into the discharge pipe and initiating flush. When water level reaches B, air enters the siphon and stops the flush.

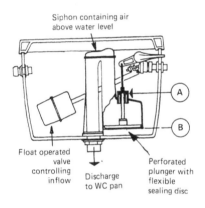

Siphon containing air above water level

Float operated valve controlling inflow

Discharge to WC pan

Perforated plunger with flexible sealing disc

A

B

Diagram 139 - Flushing cistern

Dual flush

This is an arrangement whereby air can be admitted through an aperture into the siphon and breaking flow.   Diagram 140 shows the distinction between single and dual flush design at point A on diagram 139.

The aperture admits air when the falling water level reaches it providing the lever has been released.   If the lever is held down the entry of air into the siphon is prevented.

177

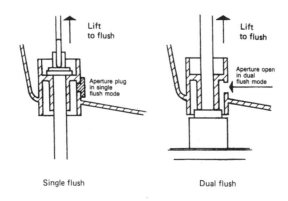

Single flush   Dual flush

Diagram 140

Automatic flushing cistern

An automatic flushing cistern (see diagram 141) contains a complex arrangement of siphons in which air is trapped preventing flow down the discharge pipe until the cistern is full. At this point a subsidiary siphon empties and trapped air is drawn out releasing flow into the discharge pipe. Flow continues until the cistern is nearly empty and air can enter the main siphon from the bottom at B. The cistern commences refilling and the cycle is automatically repeated. The frequency of operation is dependent on the rate of filling.

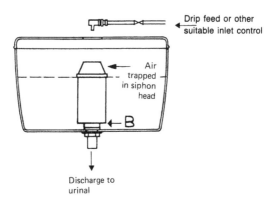

Diagram 141

Flushing troughs

Flushing troughs are used in places such as schools, factories, or where quick refill of flushing apparatus is required because of their high usage. Each watercloset is provided with a siphon in the trough fitted with a device to ensure that the volume of the flush does not exceed the prescribed limits (see diagram 142).

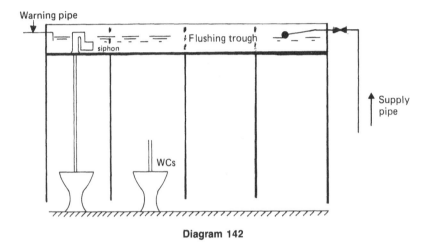

Diagram 142

BYELAW 74 - REQUIREMENTS FOR WATERCLOSET PANS

BYELAW 74. Every watercloset pan shall be -

    (a) supplied with water from a flushing cistern or trough of the valveless type which incorporates siphonic apparatus; and

    (b) so made and installed that after normal use its contents can be cleared effectively by either -

        (i) a single flush of water, or

        (ii) where the cistern or trough is designed to give flushes of different volumes, the larger or largest of those flushes.

GUIDANCE BYELAW 74(a)

Byelaw 74(a) requires that only flushing cisterns with siphonic discharge apparatus shall be used. Flushing cisterns to BS 1125 will be accepted as satisfying this byelaw.

GUIDANCE BYELAW 74(b)

Should a pan not be effectively cleansed with one flush a further flush might be required and this would lead to excessive use of water. Where the flush permitted under byelaws 75(b), 76 or 79(b) is up to 9.5 litres, the requirements of byelaw 74(b) would be satisfied by a WC pan capable of passing the cleansing test in BS 5503 Part 2 (Appendix A).

Also listed in the Directory are other WC pans that have passed the above test with a smaller flush meeting the requirements of byelaws 75(a), 76(a) and 79(a).

Diagram 143 illustrates the application of relevant parts of the byelaws to flushing cisterns other than automatic flushing cisterns. (See also byelaw 24).

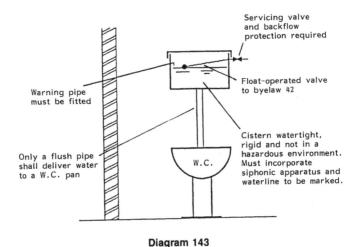

Diagram 143

BYELAWS 75 to 79
WATER USED IN CLEANSING WATER CLOSET PANS

BYELAW 75. No flushing cistern installed for use with a watercloset pan in a domestic dwelling shall give -

> (a) if it is designed or adapted to give only a single flush, a flush exceeding 7.5 litres; or
>
> (b) if it is designed or adapted to give flushes of different volumes, a maximum flush exceeding 9.5 litres.

BYELAW 76. No flushing cistern installed for use with a watercloset pan other than in a domestic dwelling shall give -

> (a) if it is designed or adapted to give only a single flush, a flush exceeding 7.5 litres, or in the case of a cistern installed before 1 January 1993 a flush exceeding 9.5 litres; or
>
> (b) if it is designed or adapted to give flushes of different volumes, and is installed before 1 January 1993, a maximum flush exceeding 9.5 litres.

BYELAW 77. Nothing in byelaws 75(a) or 76(a) shall prevent the replacement of a cistern installed before these byelaws came into operation by a similar cistern.

BYELAW 78. No cistern designed or adapted to give flushes of different volumes shall be installed after 1 January 1993 except by way of replacement of such a cistern installed before that date.

BYELAW 79. No flushing trough installed for use with two or more watercloset pans shall give -

> (a) a flush exceeding 7.5 litres; or
>
> (b) if it was installed before 1 January 1993, a flush exceeding 9.5 litres;

to any one pan.

GUIDANCE BYELAW 75 TO 79

These byelaws are intended gradually to phase out cisterns designed to give a nominal flush of 9 litres or those designed to give dual flush whether in domestic dwellings or elsewhere. By 1993 the maximum flush would become 7.5 litres (including any design tolerance) except that cisterns replacing those giving the larger or a dual flush could be of a similar capacity to the original.

Where a complete installation of a flushing cistern and pan is being replaced, the cistern will be treated as a new and not as a replacement cistern.

Until a new British Standard for flushing cisterns giving a reduced volume flush has been prepared, cisterns listed in the Directory as meeting this requirement will be acceptable as satisfying byelaw 75(a). For the remainder:

> (a) Where a cistern in Scotland is of a type to give a single flush of up to 9 litres nominal, it would be accepted if it complies with the relevant requirements of BS 1125: WC flushing cisterns (including dual flush cisterns and flush pipes.)

> (b) Where a cistern is of a type to give at the choice of the user, a flush of one or other of two volumes it would be accepted until 1 Jan 1993 if it complies with BS 1125 and it is arranged in accordance with the relevant requirements of that standard.

In the interim period before 1 Jan 1993 it will be permissible to instal in England and Wales cisterns either of 7.5 litres maximum single flush or 9.5 litres dual flush. In Scotland the byelaws vary slightly, in that until 1 Jan 1993 it will be permissible to instal cisterns giving a single flush not exceeding 9.5 litres.

BYELAWS 80 and 81

WARNING PIPES AND MARKING OF FLUSHING CISTERNS OR TROUGHS

BYELAW 80. Every flushing cistern or trough installed in any premises supplying water to a watercloset pan shall be fitted with a warning pipe and shall be indelibly marked on the inside with a line indicating the water level at which the float-operated valve is to shut off when that cistern or trough operates to comply with the relevant provision of byelaws 75 to 77 and 79.

BYELAW 81. Every flushing cistern or trough designed or adapted to give flushes of different volumes and installed in any premises shall have clearly and permanently marked on it or displayed near it instructions for operating it to obtain different flushes.

GUIDANCE BYELAW 80

The requirements of byelaws 31 to 34 and 38 to 41 do not apply to flushing cisterns so the requirement for a warning pipe is included in byelaw 80.

The marking required under byelaw 80 is related to the designed setting of the float-operated valve determining the effective capacity of the cistern and the volume of flush.

GUIDANCE BYELAW 81

Byelaw 81 requires provision of instructions to the users of dual flush cisterns. These markings must never be removed.

BYELAWS 82 and 83
WATER USED IN FLUSHING URINALS

BYELAW 82. (1) Every urinal which is cleared by water after use shall be supplied with water from a flushing cistern or flushing trough which incorporates siphonic apparatus and is designed or adapted to supply no more water than is necessary.

(2) No cistern or trough mentioned in paragraph (1) which is installed for use in connection with two or more urinal bowls or stalls or two or more widths of slab each exceeding 700mm in width shall be designed or adapted to fill at a rate exceeding -

(a) 7.5 litres per hour; or

(b) if it is installed before 1 January 1989, 15 litres per hour;

for any one urinal bowl or stall, or 700mm width of urinal slab.

(3) No cistern or trough mentioned in paragraph (1) which is installed after 1 January 1989 for use in connection with a single urinal bowl or stall shall be designed or adapted to fill at a rate exceeding 10 litres per hour.

BYELAW 83.

(1) Every pipe which supplies water to a flushing cistern or trough used for flushing a urinal shall be fitted with -

  (a) a flow shut-off device controlled by a time switch and a lockable shut-off valve; or

  (b) some other equally effective automatic device or method for regulating the periods during which the cistern or trough may fill.

(2) Paragraph (1) shall not apply to a flushing cistern or trough which is -

  (a) manually operated; or

  (b) fills and flushes by the operation of some other fitting, a signal transmitted from a photo-electric cell or pressure pad, or some other device designed to ensure that the cistern flushes only after the urinal is used.

GUIDANCE BYELAW 82

The requirements of byelaw 82(1) will be accepted as being satisfied if any flushing cistern delivering water to a urinal installation complies with BS 1876: Automatic flushing cisterns for urinals.

The frequency of operation of an automatic flushing cistern depends on the control of the rate of filling. Under byelaw 82(2) the rate of filling must be controlled to not more than 7.5 litres per hour (15 litres per hour up to the date 1 January 1989) for each urinal unit where the cistern delivers to two or more units. Where only one unit is being supplied accurate control of inflow becomes difficult and the future permitted rate will be increased to up to 10 litres per hour.

GUIDANCE BYELAW 83

In addition the owner/occupier must have means of preventing inflow during times when the urinals are not being used. This can be achieved by a time switch or by "impulse" initiated automatic control systems.

The following are three examples of installations meeting byelaw 83. Note that any combination of types of urinal and controls are permissible.

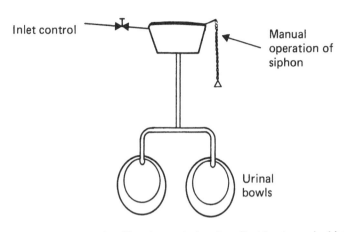

Inlet control

Manual operation of siphon

Urinal bowls

Diagram 144 - Example of hand operated system flushing two urinal bowls

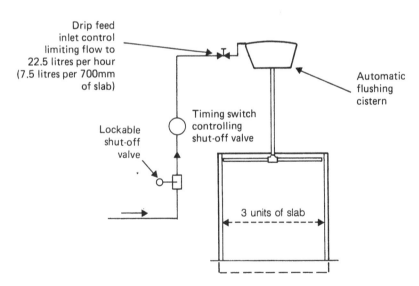

Drip feed inlet control limiting flow to 22.5 litres per hour (7.5 litres per 700mm of slab)

Automatic flushing cistern

Timing switch controlling shut-off valve

Lockable shut-off valve

3 units of slab

Diagram 145 - Example of drip feed cistern with time switch controlled flushing of a urinal slab

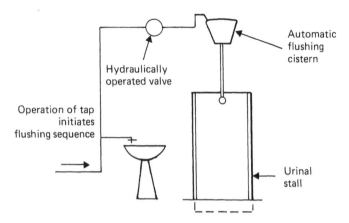

Diagram 146 - Example of control by hydraulically operated valve flushing urinal stalls

GUIDANCE BYELAWS 82 AND 83

Diagram 147 illustrates application of these byelaws.

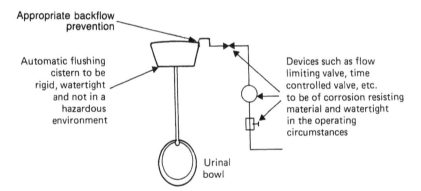

Diagram 147 - Application of relevant byelaws to automatic flushing cisterns

BYELAW 84

PIPES DISCHARGING TO WATERCLOSET PANS AND URINALS

BYELAW 84. No pipe shall be arranged to deliver water to any watercloset pan or urinal bowl, stall or slab, except -

(a) a flush pipe; or

(b) a warning pipe installed to discharge water into the air not less than 150mm above the top edge of a watercloset pan.

GUIDANCE BYELAW 84

This byelaw prevents any connection other than a flushing pipe to a watercloset pan or a urinal. It has the effect of preventing the discharge from a warning pipe to these appliances except so as to cause considerable nuisance to users. See guidance to byelaws 38 to 40.

Where there is no reasonable alternative, discharges of water from warning pipes may be taken into tundishes which, in turn, discharge water into a WC pan. There should be a clearly visible air gap between the outlet of the warning pipe and the top edge of any tundish.

If a warning pipe outlet is required to discharge via a tundish to the flush pipe, a relaxation of byelaw 84 is required by the undertakers.

PART IX

PREVENTION OF WASTE, MISUSE AND CONTAMINATION OF WATER FROM DRAW-OFF TAPS, BATHS, BASINS, SINKS AND OTHER FITTINGS

CONTENTS

PART IX - PREVENTION OF WASTE, MISUSE AND CONTAMINATION OF WATER FROM DRAW-OFF TAPS, BATHS, BASINS, SINKS AND OTHER FITTINGS

BYELAW 85

DEFINITIONS

BYELAW 85.   (1) Every bath, wash basin, sink or similar apparatus installed for use in any premises shall be -

(a) so constructed or arranged that every inlet for water is hydraulically separate from, and unconnected with, any water outlet; and

(b) provided with a watertight and readily accessible plug or some other device capable of closing the water outlet.

(2) Paragraph (1)(b) shall not apply to any

(a) shower bath or shower tray;

(b) apparatus to which water is delivered at a rate not exceeding 3.6 litres a minute, or in the case of a washing trough, 3.6 litres a minute to any unit of it, solely from a fitting designed or adapted for that purpose; or

(c) apparatus installed in any hospital or used in any medical, dental or veterinary practice which is designed or adapted for use with an unplugged outlet.

GUIDANCE BYELAW 85

Compliance with byelaw 85(1)(a) is illustrated in diagram 148.

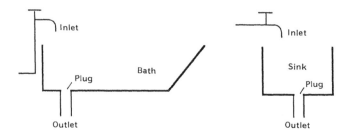

Diagram 148

The requirements of byelaw 85(1)(b) would be accepted as being satisfied whenever any bath, wash basin, sink or bidet is provided with a waste body and waste plug which comply with Tables 2 and 4 respectively of BS 3380: Wastes (excluding skeleton sink wastes) and bath overflows.

This requirement is modified by byelaw 85(2) which exempts shower baths, any apparatus (e.g. a spray tap) delivering no more than 3.6 litres per minute, and any apparatus used for medical or veterinary purposes which is designed for use with an unplugged waste. Examples are shown in Diagram 149. A tap to BS 5388: Spray taps, will be accepted as satisfying this requirement.

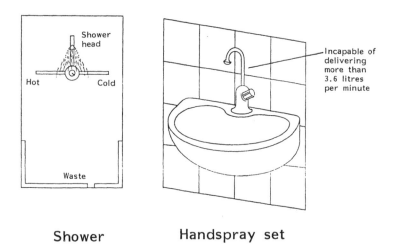

Shower   Handspray set

Diagram 149 - Appliances which are not required to be fitted with a plug

BYELAW 86
WASHING TROUGHS

BYELAW 86. Every washing trough which consists of two or more units shall be fitted with separate draw-off taps or similar apparatus for each such unit.

GUIDANCE BYELAW 86

Supply to washing troughs must be made through a fitting capable of feeding individual units without at the same time discharging to others. Examples of washing troughs are illustrated in diagrams 150 and 151.

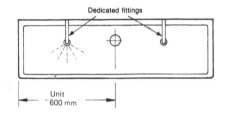

Diagram 150 - Example straight washing trough

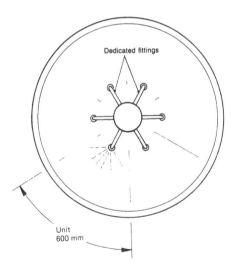

Diagram 151 - Example circular washing trough

193

BYELAW 87

REQUIREMENTS FOR DRAW-OFF TAPS

BYELAW 87. Without prejudice to byelaw 52 every draw-off tap to which water is supplied by the undertakers shall -

    (a) be capable of operating effectively at

        (i) any water temperature not exceeding 65°C, and

        (ii) any internal water pressure to which it is likely to be subject; and

    (b) be made and designed so that it may be easily closed to shut off the flow of water; and

    (c) if it incorporates a renewable seal or washer, be made or adapted so that the seal or washer can be readily renewed or replaced; and

    (d) be resistant to corrosion; and

    (e) be designed when new to withstand without leaking an internal water pressure 1.5 times that to which it will ordinarily be subject.

GUIDANCE BYELAW 87

(1) The requirements of the various parts of byelaw 87 would be accepted as being satisfied if any draw-off tap complies with one of the following British Standards:

BS 1010: Part 2 Draw-off taps and above ground stopvalves.

BS 5412: Performance of draw-off taps with metal bodies for water services.

BS 5413: Performance of draw-off taps with plastics bodies for water services.

NOTE There are no British Standards for non-concussive self closing taps. For these and other draw-off taps which do not comply with the above British Standards reference should be made to the Directory for acceptable alternatives. See also guidance to byelaw 54 regarding water hammer.

(2) The requirements of byelaw 7(1) insofar as material used in a seat washer is concerned, would be accepted as being satisfied if such complies with BS 3457: Materials for water tap and stopvalve seat washers.

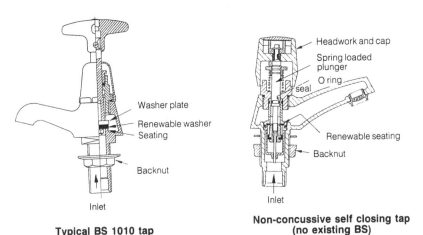

Typical BS 1010 tap

Non-concussive self closing tap
(no existing BS)

Diagram 152 - Example of taps

BYELAW 88

WATER USED BY CLOTHES AND DISHWASHING MACHINES

BYELAW 88.   (1) Except as specified in paragraph (2), no clothes washing machine, tumbler drier or dishwasher shall be connected to any supply or distributing pipe to which water is supplied by the undertakers.

(2) Paragraph (1) shall not apply to prevent the connection of

(a) a clothes washing machine which in any complete washing cycle uses not more than 3.6 litres of water for every litre of machine drum or tub volume; or

(b) a clothes washing machine incorporating a tumbler drier which in any complete washing and drying cycle uses not more than 6.4 litres of water for every litre of machine drum volume; or

(c) a tumbler drier incorporating a water spray which uses not more than 20 litres of water for every kilogram of dry load; or

(d) a dishwasher which in any complete washing cycle uses not more than 7 litres of water for every place setting.

GUIDANCE BYELAW 88

This byelaw limits the volumes of water used in a single cycle of operations of clothes and dishwashing machines and water using tumbler driers. In the latter machines the weight of dry load would be measured as cotton rating. BS 3456: Part 102: Section 102.7 specifies the type and size of material to be used in the machine for carrying out electrical safety tests, and this specification is used in consumption tests carried out by the WBAS before acceptance for listing in their Directory.

Place setting in byelaw 88(2)(d) is defined in BS 3999: Part 11 Methods of measuring the performance of household electrical appliances
- Dishwashing machines.
Reference should be made to the Directory for appliances satisfying this byelaw. See also byelaws 22 and 23.

PART X
PREVENTION OF WASTE OR CONTAMINATION OF WATER FROM ANY HOT WATER SYSTEM

CONTENTS

PART X - PREVENTION OF WASTE OR CONTAMINATION OF WATER FROM ANY HOT WATER SYSTEM

BYELAW 89

SECONDARY SYSTEM VENT PIPES

> **BYELAW 89.** No vent pipe from any secondary system shall be connected to or arranged to discharge water into any combined feed and expansion cistern connected to a primary circuit.

GUIDANCE BYELAW 89

Diagram 153 illustrates the requirements of byelaw 89.

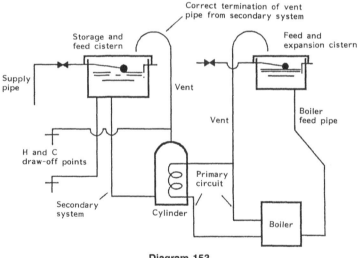

Diagram 153

BYELAW 90

ACCOMMODATION OF EXPANSION WATER IN CISTERN FED SYSTEMS

BYELAW 90. (1) Every apparatus or cylinder to which this byelaw applies shall, either -

   (a) be capable of accommodating any expansion water; or

   (b) be connected to a separate expansion cistern or vessel; or

   (c) be so arranged that expansion water can pass back through a feed pipe to any storage cistern to which that apparatus or cylinder is connected.

(2) This byelaw shall apply to any unvented apparatus or cylinder, which -

   (a) stores hot water to be drawn off for use; and

   (b) is supplied with water from a storage cistern.

GUIDANCE BYELAW 90

When water is heated it expands in volume and this expansion must be catered for because in normal operational conditions no water must be discharged to waste. Between 4°C (39°F) and 100°C (212°F) it expands by approximately 4% ($^1/_{25}$th) of its volume, e.g. an installation which holds 115 litres (25 gallons) when cold, would need to have provision built in to accommodate an extra $4^1/_2$ litres of water on heating up. This is achieved on vented hot water systems by the return of water via the feed pipe to the same feed cistern which supplies the hot water apparatus. On unvented hot water systems an expansion vessel may be employed which, due to the fact that air and gas are compressible, acts as a "cushion" to accommodate the extra volume of expanded water. Acceptable methods are illustrated in diagram 154.

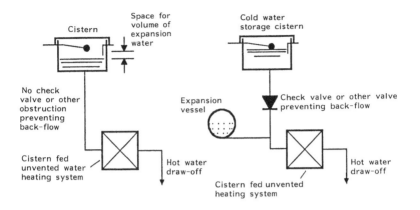

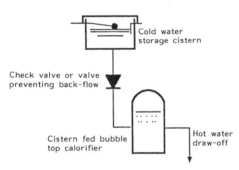

Accommodation of expansion
water in a water heater

Diagram 154 - Examples of compliance with byelaw 90 Cistern fed unvented systems

BS 6144:   Expansion vessels using an internal diaphragm, for unvented hot water supply systems, are acceptable. An example showing the operation of an expansion vessel is illustrated in diagram 155.

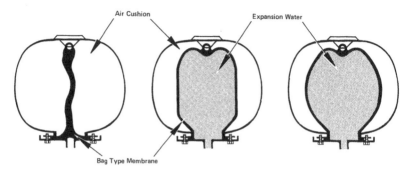

Diagram 155 - Principle of operation of air or gas loaded expansion vessels

NOTE   Cold water taps should not be connected to the feed pipe to hot water cylinders because of the risk of draining down the latter.   This is not a byelaw matter except that such an arrangement could contravene byelaw 90(1)(c) unless a check valve is fitted to the inlet of the cylinder to prevent expansion water finding its way into the cold water system.

The vessel must be sized correctly to ensure that it is capable of accepting at least 4% of the total system's water content to prevent unacceptably high pressures arising within the system.   To prevent this, careful note should be taken of the manufacturer's specification when installing these vessels.

BYELAW 91

ACCOMMODATION OF EXPANSION WATER IN SYSTEMS CONNECTED TO A SUPPLY PIPE

BYELAW 91. Every unvented water heater connected to a supply pipe and not being an instantaneous water heater shall, either -

(a) itself be capable of accommodating any expansion water; or

(b) be connected to a separate expansion cistern or vessel; or

(c) be installed so that any expansion water can be accommodated in the pipework of any secondary system provided that no hot water can enter any communication pipe or supply pipe to which a cold water draw-off tap is connected.

This byelaw ensures that installations of mains fed secondary water heating apparatus shall have a means of accommodating the expansion water. If the system pipework cannot be arranged to accommodate expansion water within itself, e.g. if a check valve or stopvalve with a loose jumper prevents this, an expansion vessel must be provided either within the secondary system itself or external to it on the cold water feed pipe. If such a vessel is included a check valve should be installed on the supply pipe upstream of the connection of that vessel not only to ensure efficient operation but also to prevent reverse flow of hot water into the supply pipe. See diagram 156.

NOTE If it is certain that no check valve, stop valve with a loose jumper, or other fitting can at any time prevent reverse flow it would be permissible that expansion water could be accommodated in an enlarged pipe or other vessel displacing only cold water back into the supply pipe.

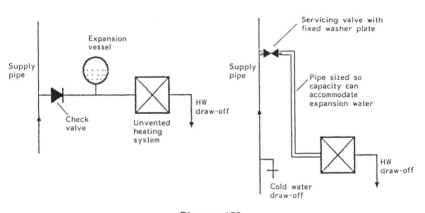

Diagram 156

NOTE   No provision is needed for expansion water in instantaneous water heaters.

BYELAW 92

CAPACITY OF EXPANSION CISTERNS OR VESSELS, ETC.

> BYELAW 92. Every expansion cistern or vessel and every cold water combined feed and expansion cistern connected to a primary circuit shall be -
>
> (a) able to accommodate any expansion water from the primary circuit to which it is connected; and
>
> (b) installed so that in ordinary operation the water level is not less than 25mm below the overflowing level of the warning pipe connected to it.

GUIDANCE BYELAW 92

In a cistern fed system expansion water in the primary circuit returns through the feed pipe into the feed and expansion cistern. The volume is approximately 4% of the total volume of water in the primary circuit. To avoid overflow the cistern capacity and the setting of the float valve must be designed to accommodate this volume. This is done by setting the level of water in the cistern when the system is cold sufficiently low so that when the system heats up the water rises to a point not higher than 25mm below the warning pipe (see diagram 157). Note that the volume of water in the cistern when the system is cold is not prescribed but it is recommended that it is not less than the capacity of the primary circuit and it must be sufficient to permit the satisfactory operation of the float operated valve.

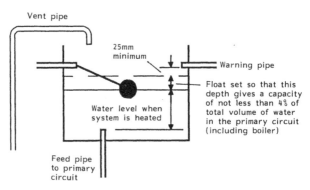

Diagram 157 - Capacity of expansion cisterns

204

Similarly in unvented systems any expansion vessel, e.g. that illustrated in diagram 155, or a cistern receiving expansion water via a distributing pipe, e.g. as shown in diagram 157, must be designed to accommodate the designed volume of expansion water without overflow or excessive pressure rise respectively.

BYELAW 93

BOILERS

> **BYELAW 93.** Every boiler shall be constructed of materials the nature, strength and thickness of which is capable of withstanding the internal water pressure and operating temperature to which it is, or is likely to be, subject.

GUIDANCE BYELAW 93

The requirements of this byelaw would be accepted as being satisfied if a boiler complies with one of the following:

BS 4433:   Solid smokeless fuel boilers with rated outputs up to 45kw. Part 1 Boilers with undergrate ash removal.

BS 4433:   Solid smokeless fuel boilers with rated outputs up to 45kw. Part 2 Gravity feed boilers designed to burn small anthracite.

BS 3377:   Boilers for use with domestic solid mineral fuel appliances.

BS 5258:   Safety of domestic gas appliances. Part 1 Central heating boilers and circulators.

BS 4876:   Performance requirements for domestic flued oil burning appliances (including test procedures).

NOTE   Subject to normal working pressure in service not exceeding the appropriate BS rating.

BYELAWS 94 and 95

PRESSURE RELIEF VALVES, EXPANSION VALVES,
TEMPERATURE RELIEF VALVES, ETC.

BYELAW 94. Every pressure relief valve, expansion valve, temperature relief valve or combined temperature and pressure relief valve connected to any boiler or hot water cylinder or storage tank or pipe shall -

    (a) close automatically after discharging water;

    (b) be watertight when closed;

    (c) be resistant to corrosion;

    (d) be constructed and installed so that the discharge of water from the valve (or from any pipe connected to it) is readily visible; and

    (e) except in the case of a temperature relief valve or the temperature function of a combined temperature and pressure relief valve, discharge water only when it is subject to water pressure 0.5 bar above the water pressure to which the boiler or other apparatus is, or is likely to be, subject.

BYELAW 95. Every unvented hot water cylinder or storage tank which is supplied with water by the undertakers (whether or not by means of a storage cistern) and which is fitted with a non-mechanical safety device shall be fitted with a temperature relief valve which -

    (a) operates or is designed to operate at a temperature not less than 5°C below that at which that safety device operates or is designed to operate;

    (b) closes automatically after discharging water; and

    (c) is watertight when closed.

GUIDANCE BYELAWS 94 AND 95

Because pressures in the undertaker's mains can vary considerably and are usually higher at night, care should be taken in selecting relief valves to ensure they are appropriate at all conditions of mains pressure.

If in doubt the undertakers should be consulted and a pressure reducing valve might be necessary. For a list of acceptable valves reference should be made to the Directory.

Note that in complying with byelaw 94(d), it is essential that any discharge should terminate at the equivalent of a Type A air gap.

The requirements of byelaw 94 would be accepted as being satisfied if every temperature relief valve complies with BS 6283: Part 2 or Part 3. See Diagram 158. Such valve incorporates easing gear for manually raising valve disc off its seating. This is required for testing that the valve is operating satisfactorily.

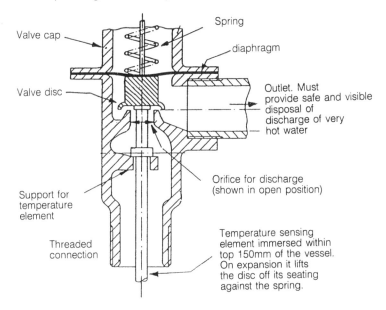

Diagram 158 - Example of a temperature relief valve

The requirements of byelaw 94 would be accepted as being satisfied if every expansion or pressure relief valve complies with BS 6283: Part 1 or Part 3 respectively. See Diagram 159.

These valves also incorporate easing gear for manually raising valve discs off their seating.

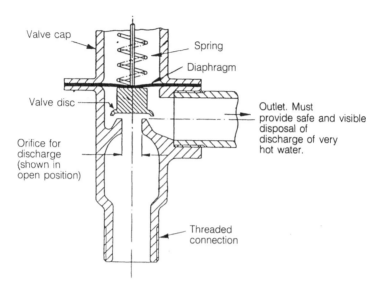

Diagram 159 - Example of a pressure relief or expansion valve

Valves must be suitable for the variable inlet pressures likely to be met in the supply area in which they are installed and must be marked clearly with their maximum pressure ratings. They must also be capable of withstanding 1.5 times the maximum pressures to which they are likely to be subjected.

Note that byelaw 95 only applies in the event of a non-mechanical safety device (e.g. a fusible plug) being fitted.

Under normal conditions of operation, expansion of water on heating up is catered for by measures dealt with in byelaws 90 to 92. In these circumstances no discharge should take place from relief valves. Only under malfunction conditions, e.g. a failure of control of heat input, should a relief valve operate and discharge water. It is therefore essential that such discharge is made in a conspicuous position and the fault remedied to prevent waste of water. The criteria which apply to warning pipes are useful guides as to what is acceptable but care should be taken to ensure safe disposal of what could be very hot water.

PART XI
TAPS FOR DRAWING
DRINKING WATER

PART XI
TAPS FOR DRAWING DRINKING WATER

BYELAW 96
TAPS FOR DRINKING WATER

BYELAW 96. (1) In every premises to which this byelaw applies, a draw-off tap convenient for drawing drinking water shall be connected to -

(a) a service pipe; or

(b) a pump delivery pipe drawing water from a service pipe; or

(c) distributing pipe drawing water exclusively from a storage cistern which is -

(i) installed in accordance with byelaw 30, and

(ii) supplied with water from a service pipe, or a pump delivery pipe drawing water from a service pipe.

(2) This byelaw applies to every premises to which the undertakers supply a separately chargeable supply of water for domestic purposes, not being premises to which section 57 of the Factories Act 1961 or section 11 of the Offices, Shops and Railway Premises Act 1963 apply.

GUIDANCE BYELAW 96

In every dwelling a cold water supply directly from the supply pipe to a conveniently situated draw-off tap shall be provided. Such a tap is that provided at a kitchen sink.

In an installation in a new single dwelling, new commercial or industrial premises, and wherever else reasonably practicable, a pipe leading to a mains drinking water tap shall be connected directly to the supply pipe prior to any water softener.

PART XII
NOTICES TO
UNDERTAKERS

PART XII
NOTICES TO UNDERTAKERS
BYELAWS 97 and 98

NOTICES TO UNDERTAKERS

BYELAW 97. (1) A person who, in respect of any premises to which water is supplied by the undertakers, proposes to carry out relevant work mentioned in paragraph (2) shall give written notice to the undertakers not less than five working days before he commences, or proposes to commence, that work.

(2) Relevant work for the purpose of paragraph (1) means the installation or alteration (other than by repair or renewal) of any -

(a) bidet; or

(b) flushing cistern; or

(c) hose union tap or tap to which a hose may be connected; or

(d) water fitting which, if there is a backflow or backsiphonage of water through it, will or may contaminate water supplied by the undertakers.

BYELAW 98. A person who, in respect of any pipe which conveys, or is intended to convey, water supplied by the undertakers, proposes to -

(a) backfill any excavation in which it is laid; or

(b) thread it through any tubular duct which enters a building below ground level; or

(c) embed it in any solid floor or wall; or

(d) lay it underground by means of a mole plough or similar apparatus,

shall give written notice to the undertakers not less than five working days before he commences, or proposes to commence, that work.

GUIDANCE BYELAWS 97 AND 98

There are variations to these byelaws in Scotland as follows :-

"For Byelaw 97 there shall be substituted -

97. A person who, in respect of any premises to which water is supplied by the undertakers, proposes to carry out any installation or alteration (other than in the way of repair or renewal) of any water fitting shall give written notice to the undertakers not less than five working days before he commences or proposes to commence that work.".

For Byelaw 99 there shall be substituted -

99. A person contravening any of these byelaws shalll be liable on summary conviction to a fine not exceeding -

(a)   level 4 on the standard scale in respect of each offence, and

(b)   in respect of a continuing offence, a further fine not exceeding £50 for each day during which the offence continues after conviction.".

PART XIII
PENALTIES

BYELAW 99

PENALTIES

BYELAW 99. Any person contravening any of these byelaws shall be liable on summary conviction to a fine not exceeding

    (a) £400 in respect of each offence; and

    (b) £40 in respect of each continuing offence, for each day during which the offence continues after conviction.

BYELAW 100

DEFENCE FOR PERSON CHARGED WITH AN OFFENCE

BYELAW 100. It shall be a defence for a person charged with an offence under these byelaws to show that -

    (a) he took all reasonable steps and exercised due diligence to avoid commission of that offence; or

    (b) he had a reasonable excuse for his act or failure to act.

GUIDANCE BYELAWS 99 AND 100

"Where work - being the supply and connection or alteration or disconnection of a water fitting - is or has been carried out by, or under the direction of, a person who participates in a certification scheme operated by a local water undertaking the consumer would not be considered as having contravened any requirements of byelaw 2 or 3(a)."

Failure to maintain, repair or replace damaged or worn fittings or components likely to cause waste or contamination of water could contravene byelaw 3(b).

PART XIV
REVOCATION

BYELAW 101

REVOCATION

101. The byelaws made by the undertakers under
.... on .... and confirmed by the
.... on .... are revoked.

APPENDICES

CONTENTS

APPENDIX A - CONSTITUTION OF PANEL PRODUCING GUIDANCE

Members

Mr N A Thompson	-	Severn Trent Water Authority (Chairman)
Mr A Aston	-	British Plumbing Fittings Manufacturers Association
Mr C Forrester	-	Strathclyde Regional Council
Mr F Howarth	-	Wessex Water Authority
Mr R J Howse	-	British Plastics Federation
Mr E G Moss	-	North Surrey Water Company
Mr C Robertshaw	-	Water Training
Mr M A Rymill	-	British Bathroom Council
Mr A Watts	-	Institute of Plumbing
Mr G D Mays	-	WRc Water Byelaws Advisory Service (Secretary)

Assisted by

Mr J Halford	-	WRc Water Byelaws Advisory Service
Mr J A Jones	-	WRc Water Byelaws Advisory Service
Mrs J N Whitaker	-	Water Research Centre
Mr S F White	-	Water Research Centre

APPENDIX B - CONVERSION TABLES

LENGTH	
Metric	**Approximate Imperial equivalent**
5mm	$3/16''$
10mm	$3/8''$
15mm	$5/8''$
20mm	$3/4''$
25mm	$1''$
40mm	$1\ 5/8''$
75mm	$3''$
100mm	$4''$
150mm	$6''$
300mm	$12''$
500mm	$1'\ 8''$
750mm	$2'\ 6''$
1 metre	$3'\ 3''$
1.35 metres	$4'\ 6''$
2 metres	$6'\ 7''$
2.5 metres	$8'\ 3''$
5 metres	$16'\ 6''$

VOLUME	
Metric litre	**Approximate Imperial equivalent**
1.0	0.22 gallons
3.6	0.79 gallons
6.4	1.4 gallons
7.0	1.5 gallons
9.6	2.1 gallons
10.0	2.2 gallons
18.0	4 gallons
20.0	4.4 gallons
50.0	11 gallons
100.0	22 gallons
500.0	110 gallons
1 000.0	220 gallons
5 000.0	1100 gallons
10 000.0	2200 gallons

RATE OF FLOW	
Metric	**Approximate equivalent gallons/minute**
7.5 litres/hour	0.0275
10 litres/hour	0.037
15 litres/hour	0.055
3.6 litres/min	0.8
10 litres/min	2.2

PRESSURE (approximate equivalents)		
Bars	**lbs/sq in**	**Fead head water**
0.005	0.07	$2''$
0.010	0.145	$4''$
0.05	0.72	$1'\ 8''$
0.10	1.45	$3'\ 4''$
0.50	7.25	$17'$
1.00	14.5	33.5
3.00	43.5	100
6.00	87	200
10.00	145	335
12.00	174	400
16.00	230	535
25.00	360	840

NOTE: a millibar is 0.001 of a bar

APPENDIX C - LIST OF BRITISH STANDARDS CITED IN GUIDE

BS Number, Title and Guide Reference

21 Pipe threads for tubes and fittings where pressure tight joints are made on the threads (metric)   52 (GS 2)

143 and 1256 Malleable cast iron and cast copper alloy threaded pipe fittings   52 (GS 2)

417 Part 2: Galvanised low carbon steel cisterns, cistern lids, tanks and cylinders (metric) 52 (GS 11)

486 Asbestos cement pressure pipes and joints 52 (GS 5)

534 Steel pipes and specials for water and sewage 52 (GS 2)

699 Copper direct cylinders for domestic purposes 52 (GS 11)

843 Thermal storage electric water heaters (constructional and water requirements) 52 (GS 11)

853 Calorifiers and storage vessels for central heating and hot water supply 52 (GS 11)

864 Part 2: Capillary and compression fittings for copper tubes 52 (GS 4, 10)
Part 3: Compression fittings for polyethylene pipes 52 (GS 9)
Part 5: Compression fittings of copper and copper alloy for polyethylene pipes 52 (GS 7)

1010 Part 2: Draw-off taps and above ground stopvalves 16, 64, 87

1125 WC flushing cisterns (including dual flush cisterns and flush pipes) 52 (GS 11, 75 to 79)

1188 Ceramic wash basins and pedestals 16

1189 Baths made from porcelain enamelled cast iron 16

1212 Float operated valves (excluding floats)
Part 1: Piston type 42
Part 2: Diaphragm type (brass body) 12 (GS Gen)
Part 3: Diaphragm type (plastics body) for cold water services 24 (2), 25 (GS Gen), 42

1244 Metal sinks for domestic purposes
Part 2: Stainless steel sink tops 16

1387 Screwed and socketted steel tubes and tubulars for plain and steel tubes suitable for welding or for screwing to BS 21 pipe threads 52 (GS 2)

1390 Sheet steel baths 16

1475 Wrought aluminium and aluminium alloy for general engineering purposes - wire 52 (GS 11)

1563 Cast iron sectional tanks (rectangular) 52 (GS 11)

1564 Pressed steel sectional rectangular tanks 52 (GS 11)

1565 Part 2: Galvanised mild steel indirect cylinders, annular or saddleback type.   Metric units 52 (GS 11)

1566 Copper indirect cylinders for domestic purposes
Part 1: Double feed indirect cylinders 36, 52 (GS 11)
Part 2: Single feed indirect cylinders 37, 52 (GS 11)

1710 Identification of pipelines and services 27

1724 Bronze welding by gas 52 (GS 4)

1740 Wrought steel pipe fittings (screwed BS 21 - R series thread) 52 (GS 2)

1876 Automatic flushing cisterns for urinals 52 (GS 11), 82

1965 Butt welding pipe fittings for pressure purposes
Part 1: Carbon steel 52 (GS 2)

1968 Floats for ball valves (copper) 42

1972 Polythene pipe (Type 32) for above ground use for cold water services 52 (GS 7)

2456 Floats (plastics) for ball valves for hot and cold water 42

2494 Elastomeric joint rings for pipework and pipelines 52 (GS 2, 3)

2580 Underground plug cocks for cold water services 64

2594 Carbon steel welded horizontal cylindrical storage tanks 52 (GS 11)

2871 Copper and copper alloys, Tubes

Part 1: Copper tubes for water, gas and sanitation 52 (GS 4)

2872 Copper and copper alloys.   Forging stock and forgings 52 (GS 4)

2874 Copper and copper alloy rods and sections (other than forging stock) 52 (GS 4)

2879 Draining taps (screw down pattern) 66

3198 Copper hot water storage combination units for domestic purposes 30 (2), 52 (GS 11)

3377 Boilers for use with domestic solid mineral fuel appliances 93

3380 Wastes (excluding skeleton sink wastes) and bath overflows 85

3416 Black bitumen coating solutions for cold applications 52 (GS 11)

3445 Fixed agricultural water troughs and water fittings 12 (GS 4), 46

3456 Part 102.7: Safety of household and similar electrical appliances 88

3457 Materials for water tap and stopvalve seat washers 87

3505 Unplasticised polyvinyl chloride (PVC-U) pressure pipes for cold potable water 52 (GS 6)

3999 Methods of measuring the performance of household electrical appliances.
Part 11: Dishwashing machines 88

4127 Light gauge stainless steel tubes
Part 2: Metric units 52 (GS 10)

4213 Cold water storage and feed and expansion cisterns, (polyolefin or olefin copolymer) and cistern lids 52 (GS 11)

4305 Baths for domestic purposes made from cast acrylic sheets 16

4346 Joints and fittings for use with unplasticised PVC pressure pipes 52 (GS 6)

4433 Solid smokeless fuel boilers with rated outputs up to 45kW
Part 1: Boilers with undergrate ash removal 93
Part 2: Gravity fed boilers designed to burn small anthracite 93

4504 Flanges and bolting for pipes, valves and fittings, metric series
Part 1: Ferrous 52 (GS 2, 3)
Part 2: Copper alloy and composite flanges 52 (GS 4)

4622 Grey iron pipes and fittings 52 (GS 3)

4772 Ductile iron pipes and fittings 52 (GS 3)

4876 Performance requirements for domestic flued oil burning appliances (including test procedures) 93

4991 Propylene copolymer pressure pipe (Series 1) 52 (GS 8)

5114 Performance requirements for joints and compression fittings for use with polyethylene pipes 52 (GS 9)

5163 Predominantly key-operated cast iron gate valves for waterworks purposes 64

5258 Safety of domestic gas appliances
Part 1: Central heating boilers and circulators 93

5337 Code of practice for the structural use of concrete for retaining aqueous liquids 30(1), 47

5388 Spray taps 85

5412 Performance of draw-off taps with metal bodies for water services 16, 87

5413 Performance of draw-off taps with plastics bodies for water services 16, 87

5433 Underground stopvalves for water services 64

5503 Part 2: Vitreous china washdown WC pans with horizontal outlet 74

5505 Bidets Part 3: Vitreous china bidets, over-rim supply only 19 (GS 1)

5506 Washbasins Part 3: Washbasins (one or three tap holes) 16

5556 General requirements for dimensions and pressure ratings for pipe of thermoplastic materials 52 (GS 7)

5750 Quality systems 2, 3

6076 Tubular polyethylene film for use as protective sleeving for buried iron pipes and fittings 52 (GS 3)

6144 Expansion vessels using an internal diaphragm, for unvented hot water supply systems 90

6280 Method of vacuum (backsiphonage) test for water using appliances 11 (GS 2), 24

6281 Devices without moving parts for the prevention of contamination of water by backflow
Part 1: Type A air gaps 11 (GS 1)
Part 2: Type B air gaps 11 (GS 2)
Part 3: Pipe interrupters of nominal size up to and including DN 42 11 (GS 5)

6282 Devices with moving parts for the prevention of contamination of water by backflow
Part 1: check valves of nominal size up to and including DN 54 11 (GS 3), 72
Part 3: In-line anti-vacuum valves of nominal size up to and including DN 42 11 (GS 6), 72
Part 4: Combined check and anti-vacuum valves of nominal size up to and including DN 42 11 (GS 7), 72

6283 Safety devices for use in hot water systems
Part 1: Expansion valves for pressures up to and including 10 bar 94
Part 2: Temperature relief valves for pressures up to and including 10 bar 94
Part 3: Combined temperature and pressure relief valves for pressures up to and including 10 bar 94

6572 Blue polyethylene pipe up to nominal size 63 for below ground use for potable water 52 (GS 7)

6614 Safety devices and water supply connections for washing machines and dishwashers connected to the water supply mains 22

6675 Servicing valves (copper alloy) for water services 71

6700 Design, installation, testing and maintenance of services supplying water for domestic use within buildings and their curtilages Notes 5, 39, 49, 52 (GS 1), 53/54

6730 Black polyethylene pipes up to nominal size 63 for above ground use for cold potable water

6920 Suitability of non-metallic products for use in contact with water intended for human consumption with regard to their effect on the quality of water 7

CP 312 Plastics pipes (thermoplastics materials) Parts 1 to 3 52 (GS 1)

CP 2010 Pipelines Parts 1 to 5 52 (GS 1)

GUIDE REFERENCES

In this index references to the byelaw number also refer to the relevant guidance.   In some instances additional references are given to sections of guidance under the quoted byelaw, such references are prefixed GS (Guidance Section).

References are also made to "Notes" and "Foreword" which can be found at the beginning of the volume.   References to Schedules in byelaw 25 are denoted "Sch A", "Sch B" or "Sch C".

NOTES

NOTES